由重庆工商大学资助出版

U0663871

Carbon nitride based semiconductors for photocatalytic hydrogen evolution from water splitting

氮化碳基半导体光解水制氢

— 英文版 —

刘学成　姚斌　祁凝　许晗宇　著

化 学 工 业 出 版 社

·北 京·

内容简介

This book systematically explores carbon nitride-based semiconductors for photocatalytic hydrogen production through water splitting. Beginning with fundamental concepts of hydrogen generation, it details modifications to pure g-C_3N_4 and interfacial engineering designs before examining various composite materials: CeO_2/g-C_3N_4 nanocomposites demonstrate enhanced charge separation, while CoO nanoparticle integration improves visible-light absorption. The text analyzes N-doped ZnO/g-C_3N_4 heterojunctions, MnO_2-loaded architectures for oxidative stability, and oxygen-deficient $LaVO_4$/g-C_3N_4 systems. Advanced configurations include Co-C_3N_4/$BiPO_4$ dual-cocatalysts, atomically dispersed Co-N_4 sites in 2D frameworks, and B/P-doped variants for bandgap modulation, concluding with molten-salt synthesized Fe@C_3N_4 nanosheets. Each system's synthesis, characterization, mechanistic pathways, and hydrogen evolution performance are rigorously evaluated, culminating in forward-looking perspectives for next-generation photocatalyst development.

图书在版编目（CIP）数据

氮化碳基半导体光解水制氢 = Carbon nitride based semiconductors for photocatalytic hydrogen evolution from water splitting : 英文 / 刘学成等著.

北京 ： 化学工业出版社，2025.8. -- ISBN 978-7-122 -48746-9

Ⅰ．TE624.4

中国国家版本馆 CIP 数据核字第 20257ZC018 号

责任编辑：高　宁
责任校对：宋　玮　　　　　　　装帧设计：韩　飞

出版发行：化学工业出版社
　　　　　（北京市东城区青年湖南街 13 号　邮政编码 100011）
印　　装：北京建宏印刷有限公司
710mm×1000mm　1/16　印张 11¾　字数 300 千字
2025 年 8 月北京第 1 版第 1 次印刷

购书咨询：010-64518888　　　售后服务：010-64518899
网　　址：http://www.cip.com.cn
凡购买本书，如有缺损质量问题，本社销售中心负责调换。

前　言

如今，全球环境问题和能源危机越来越引起人们的高度重视。众所周知，煤炭和石油是不可再生能源，随着社会的高速发展，石油的开采量也日益增加，从而导致全球的石油储存量大幅度减少，我国面临着能源转型的挑战。氢能，不仅具有良好的可再生性和燃烧性等优异特点，而且其来源广泛，可以大量储存。利用光催化技术分解水产生氢气，既不会产生二次污染，又可以通过半导体催化剂使太阳能转化为氢能。

作为非金属半导体催化材料，氮化碳在光催化制氢技术领域具备极好的潜在发展前景和商业化应用前景，但未经改性的氮化碳存在比表面积小、光生载流子（电子和空穴）分离和迁移速率慢、光吸收范围较窄、电子-空穴对容易复合等缺点，导致其光催化分解水产氢的性能不理想。本文针对氮化碳光吸收利用率低和光生电子与空穴复合率高等关键问题，通过对氮化碳掺杂改性和构建异质结等方法来提高光解水产氢的效率，并分析了氮化碳基材料的光解水产氢机理。

本书凝聚了笔者研究团队的最新成果，系统阐述了氮化碳基光催化材料的制备方法、改性策略及其在光解水制氢中的应用。希望能够为从事光催化分解水制氢研究人员提供较为详尽的氮化碳基材料的制备技术及其在光催化分解水的应用，为光催化分解水的研究提供一定科学依据和技术支撑。

全书共 10 章，刘学成负责撰写第 1~3 章（10 万字），姚斌负责撰写第 4~6 章（10 万字），祁凝负责撰写第 7~8 章（5 万字），许晗宇负责撰写第 9~10 章（5 万字）。感谢重庆工商大学学术著作出版基金资助。全书由刘学成统稿。由于研究领域的快速发展及编者水平所限，书中难免存在疏漏之处，恳请学界同仁不吝指正。

著者

Preface

Nowadays, global environmental issues and energy crises are attracting significant attention worldwide. As fundamental non-renewable energy sources, coal and oil face depleting reserves due to accelerated extraction rates driven by rapid societal development. In this context, China faces growing challenges in energy transition. Hydrogen energy emerges as a promising solution, possessing high renewability potential, exceptional combustibility, abundant availability, and large-scale storage capacity. Particularly, photocatalytic water splitting technology enables solar-to-hydrogen energy conversion through semiconductor catalysts without generating secondary pollution.

Among non-metallic semiconductor materials, graphitic carbon nitride (g-C_3N_4) demonstrates significant potential for photocatalytic hydrogen production. However, pure carbon nitride exhibits inherent limitations including limited specific surface area, inefficient photoinduced charge carrier migration, narrow spectral absorption range, and rapid electron-hole recombination, resulting in suboptimal photocatalytic hydrogen production efficiency. This monograph focuses on enhancing light absorption/utilization efficiency and suppressing charge recombination in carbon nitride through strategic doping modification and heterojunction construction, while systematically analyzing the photocatalytic mechanisms of carbon nitride-based materials.

This publication aims to provide researchers in photocatalytic water splitting with comprehensive preparation methodologies and application guidelines for carbon nitride-based materials, offering scientific foundation and technical references for advancing this field. The content synthesizes extensive research literature, with substantive contributions from multiple authors: Liu Xuecheng (Chapters 1-3, 100,000 words), Yao Bin (Chapters 4-6, 100,000 words), Qi Ning (Chapters 7-8, 50,000 words), and Xu Hanyu (Chapters 9-10, 50,000 words).

We gratefully acknowledge the academic publication fund support from Chongqing Technology and Business University. The monograph was revised and compiled by Liu Xuecheng. Due to the limited level of the editor, there may be inappropriate content selection and wording in the book. We kindly ask readers and peers to criticize and correct us.

Authors

Contents

Chapter 1

Introduction

1.1 Hydrogen evolution

Since human are becoming more conscious of the need to protect the environment and conserve energy, the global energy crisis and demand have greatly increased research into green and renewable energy. It is generally recognized that hydrogen (H_2) is a promising clean energy because it has the advantages in releasing environmental contamination and fossil fuel energy crisis. The chemical energy stored in the H—H bond is easily released when oxygen is combined with it, forming only water vapor as the reaction byproduct. Photocatalytic water splitting to form hydrogen has been a hot research area in recent years. Water electrolysis using solar light replicates photosynthesis by converting water into hydrogen (H_2) and oxygen (O_2) with the aid of inorganic photo-semiconductors that catalyze the water-splitting reaction. Hydrogen is widely considered to be the future clean energy carrier in many applications of environmentally friendly vehicles, domestic heating, and stationary power generation, and so on.

In current, large-scale hydrogen generation is performed from fossil fuels. Numbers of research groups are focused on developing efficient semiconductor materials for photocatalytic water splitting to generate hydrogen.

An essential prerequisite for the photocatalyst is its resistance to reaction at the solid/liquid interface to comprise its properties. However, it is difficult to find an ideal photocatalyst to meet all the requirements (chemical stability, corrosion resistance, visible light harvesting and suitable band edges) that would render photocatalytic H_2 production a viable alternative. Fortunately, nano science and nanotechnology have boosted the modification of existing photocatalysts and the discovery and development of new candidate materials. The rapidly increasing number of scientific publications constitutes clear bibliographical evidence for the significance of this hot topic. Since 2004, the number of publications on nano photocatalytic hydrogen generation has increased by a factor of about 1.5 times every year. Many papers have examined the impact of different nanostructures and nanomaterials on the performance of photocatalysts, since their energy conversion efficiency is principally determined by nano-scale properties. To tackle these challenges, it is highly desirable to search for novel visible-light-driven semiconductor materials and further fabricate highly efficient systems/architectures for energy supply and environmental remediation.

The graphite-like carbon nitride (g-C$_3$N$_4$), as a metal-free polymer n-type semiconductor, possesses many promising properties, such as unique electric, optical, structural and physiochemical properties, which makes g-C$_3$N$_4$-based materials a new class of multifunctional nanoplatforms for electronic, catalytic and energy applications. Especially, g-C$_3$N$_4$-based photocatalysts have attracted increasing interest worldwide.

Thus, the g-C$_3$N$_4$-based nanostructures are emerging as ideal candidates for a variety of energy and environmental photocatalytic applications, such as photocatalytic water reduction and oxidation, degradation of pollutants and carbon dioxide reduction. More interestingly, although the annual number of publications about g-C$_3$N$_4$-based photocatalysts is much smaller than that about TiO$_2$ photocatalysts, the publications of g-C$_3$N$_4$ photocatalysis present an obvious approach to those of graphene-based photocatalysis. Furthermore, g-C$_3$N$_4$-based photocatalysts along with graphene-based ones, have significantly reduced the proportion of TiO$_2$ photocatalysts. Notably, g-C$_3$N$_4$, as a highly promising sustainable material, has emerged as a prominent candidate in the photocatalytic field.

1.2 Modification of pure g-C$_3$N$_4$

The g-C$_3$N$_4$ can absorb solar light with a wavelength of less than 460 nm and has a bandgap of ~2.70 eV. To produce enough electrons and holes with strong driving forces to sustain the photocatalytic redox activities, a semiconductor's ideal bandgap should be 2.0 eV, which captures the whole solar visible light spectrum.

The design and development of g-C$_3$N$_4$ photocatalysis has resulted many research results to date, but the system's low efficiency keeps it far from being used in real-world scenarios. Therefore, there is little doubt that the explosive growth of g-C$_3$N$_4$-based composite photocatalysts will continue to accelerate in the near future. Consequently, an extensive range of band-gap engineering methods are utilized and have effectively demonstrated the enhancement of g-C$_3$N$_4$'s photocatalytic activity. Solid solution, molecular-level doping (copolymerization), and atomic-level doping (foreign metal in cavity, interlayer, and nonmetal elements) are some of these prominent methods.

Doping is thought to depend on the bandgap, type, composition, and shape of the photocatalyst as well as the concentration or type of electrolyte on the durability and HER activity. Understanding the impact of dopants on the physicochemical properties of g-C$_3$N$_4$ nanostructures as well as their catalytic/photocatalytic activities and mechanisms could be achieved by combining both experimental and theoretical calculations.

The electronic, physicochemical, optical, and photocatalytic properties of the g-C$_3$N$_4$ network are enhanced by the additional binding functions or doping, introduced by imported metallic impurities—either alkali or transition metals—which decrease the band gap and increase visible light absorption. Two possible forms of metal (cation) doping have been identified: cavity doping and intralayer and interlayer doping[1]. The so-called cavity doping can be achieved by incorporating the metal ions (Mn$^+$) into the large cavities, specifically the triangular pores between the triazine units of the g-C$_3$N$_4$ network in the plane of g-C$_3$N$_4$ through the strong coordination bonds between the metal and negatively charged nitrogen atoms.

It has been suggested that the large cavities of g-C$_3$N$_4$ may be doped with transition metal ions such as Fe^{3+}, Zn^{2+}, Mn^{3+}, Co^{3+}, Ni^{3+}, and Cu^{2+}. The DFT calculations for the cavity-doped Pt and Pd atoms in the g-C$_3$N$_4$ lattice were demonstrated by Pan et al.[2]. This improved carrier mobility with a narrower bandgap enhances their light harvesting power to assist various photocatalytic activities. Doped alkali-metal ions (Li$^+$, Na$^+$, and K$^+$) were placed in various intercalated regions of the g-C$_3$N$_4$ network, as reported by Xiong et al.[3]. It is reported that the intralayer non-metal doping occurs at the substituted single-layer graphitic sites of g-C$_3$N$_4$[4]. These effects included altered spatial charge distribution, increased free charge-carrier concentration, improved rate of charge-transport, and enhanced photocarrier separation. The K$^+$ doping in the g-C$_3$N$_4$ lattice,which could lower the valence band (VB) level and promote the immigration of photo-electron and photo-hole carriers under solar light was also demonstrated by Zhang et al.[5]. Alkali, transition, noble, or rare earth metals can all be used to dope the g-C$_3$N$_4$

crystals. To improve the efficiency of charge-carrier transfer, transportation, and separation to bring about the spatial charge-carrier distribution, alkali metals (K^+ and Na^+) were integrated into the nitrogen pockets of the g-C_3N_4 lattice. This led to the increased photocatalytic redox processes. The alkali concentration might be used to adjust the conduction band (CB) and VB potential positions of alkali metal-doped g-C_3N_4. While Na atoms could be positioned within the conjugated network of g-C_3N_4 lattice[6, 7], K atoms tended to exist at g-C_3N_4 interlayer positions via bridging. The K atoms formed channels and bridging layers for charge delivery, transfer, and separation of photogenerated electron-hole pairs through their chemical interlayer bonds. In contrast, Na atoms were connected to the in-plane N atom to form ionic bonds by forfeiting the 3-electrons. The optical and electrical properties of metal-doped g-C_3N_4 materials mentioned above, could also be changed by the doping of metals including Pd, Fe, Cu, W, and Zr[8-12]. By lowering the band gap, increasing charge mobility, and extending the lifetime of charge carriers, metal-doping to g-C_3N_4 can effectively boost the light absorption power and provide visible photocatalytic activity[13,14]. Due to the lone pairs of electrons in their nitrogen pots, the trapped metal cations interact strongly with the negatively charged nitrogen atoms in g-C_3N_4[15]. Noble metals like Ag, Au, Pt, and Pd were used to functionalize the g-C_3N_4 by creating the additional energy levels that prolonged the visible-light sensitivity by improving carrier mobility and electron-hole separation rate, and decreased the electron-hole charges recombination rate. In addition to the many benefits of metal-doped g-C_3N_4, Yuan et al. also listed some of its drawbacks, including its poor thermal stability and the possibility that the newly formed energy bands could operate as recombination centers by lowering quantum efficiencies[16].

Fe-doped carbon nitride is a potential photocatalyst for the decomposition of simulated seawater to produce hydrogen or oxygen and for the photocatalytic oxidation of glucosamine hydrochloride under visible light. The doping of Fe was confirmed by a series of characterizations. The symmetric π-conjugated ring of g-C_3N_4 could be disrupted by Fe doping, which could enhance the visible light absorbance and $n \rightarrow \pi^*$ electron leaps. Notably, the 0.5MCN-Fe catalyst exhibited excellent H_2 and O_2 evolution rates in the visible light of 1.41 mmol/(h·g) H_2 and 0.12 mmol/(h·g) O_2, which were about 5.54 and 7.6 times higher than those of BCN, respectively. Remarkably, the 0.5MCN-Fe catalyst could remove glucosamine hydrochloride completely by oxidation to generate 15.84 μmol/(h·g) H_2. The characterization results show that Fe-doped carbon nitride reduces the band gap energy and increases the rate of photo capture and separation of electron-hole carriers, thus improving the photocatalytic activity.

Critical barriers in g-C$_3$N$_4$ for hydrogen evolution reactions (HERs) include their low current densities, relatively high overpotential, and poor long-term stability, which prevent them from meeting realistic requirements. Noble metal dopants could be used to increase solar light harvesting and charge carrier separation during g-C$_3$N$_4$ capacity to solve this. To prevent unplanned incidents in which an H$_2$ tank explodes, the safety of H$_2$ storage tanks is also a major concern. The density could then be increased by cold or cryo-compressed hydrogen storage and the use of innovative, long-lasting, inexpensive materials for H$_2$ adsorption, which might make H$_2$ storage safer than tanks. The g-C$_3$N$_4$ should be synthesized in situ on solid carbon-cloth sheets or metal hydroxide/oxide substrate, which could serve as the cathode for the HER.

To achieve high surface area and exceptional catalytic properties, dopants can be used with different metals. In this book, the efficiency of photocatalytic hydrogen evolution of g-C$_3$N$_4$ with noble cocatalyst Pt is improved by doping, modifying or constructing heterojunction to solve the key problems such as low light absorption and high recombination rate of photogenerated electrons and holes for. And the mechanism of photocatalytic hydrogen evolution on g-C$_3$N$_4$ based material is analyzed. In the future, there may be more opportunities to investigate multiple atom co-doping g-C$_3$N$_4$ systems.

Reference

[1] Cao S., Low J., Yu J., et al. Polymeric photocatalysts based on graphitic carbon nitride[J]. Advanced Materials, 2015, 27(13): 2150-2176.

[2] Pan H., Zhang Y W., Shenoy V B., et al. Ab initio study on a novel photocatalyst: functionalized graphitic carbon nitride nanotube[J]. Acs Catalysis, 2011, 1(2): 99-104.

[3] Xiong T., Cen W., Zhang Y., et al. Bridging the g-C$_3$N$_4$ interlayers for enhanced photocatalysis[J]. Acs Catalysis, 2016, 6(4): 2462-2472.

[4] Wen J., Xie J., Chen X., et al. A review on g-C$_3$N$_4$-based photocatalysts[J]. Applied surface science, 2017, 391: 72-123.

[5] Zhang M., Bai X., Liu D., et al. Enhanced catalytic activity of potassium-doped graphitic carbon nitride induced by lower valence position[J]. Applied Catalysis B: Environmental, 2015, 164: 77-81.

[6] Zhan J., Hu S., & Wang Y. A convenient method to prepare a novel alkali metal sodium doped carbon nitride photocatalyst with a tunable band structure[J]. RSC advances, 2014, 4(108): 62912-62919.

[7] Hu S., Li F., Fan Z., et al. Band gap-tunable potassium doped graphitic carbon nitride with enhanced mineralization ability[J]. Dalton Transactions, 2015, 44(3): 1084-1092.

[8] Tonda S., Kumar S., Kandula S., et al. Fe-doped and-mediated graphitic carbon nitride nanosheets for enhanced photocatalytic performance under natural sunlight[J]. Journal of Materials Chemistry A, 2014, 2(19): 6772-6780.

[9] Li Z., Kong C., Lu G. Visible photocatalytic water splitting and photocatalytic two-electron oxygen formation over Cu-and Fe-doped g-C_3N_4[J]. The Journal of Physical Chemistry C, 2016, 120(1): 56-63.

[10] Ding J., Wang L., Liu Q., et al. Remarkable enhancement in visible-light absorption and electron transfer of carbon nitride nanosheets with 1% tungstate dopant[J]. Applied Catalysis B: Environmental, 2015, 176: 91-98.

[11] Wang, Y., Wang, Y., Li, Y., et al. Simple synthesis of Zr-doped graphitic carbon nitride towards enhanced photocatalytic performance under simulated solar light irradiation[J]. Catalysis Communications, 2015, 72: 24-28.

[12] Hu S. W., Yang L. W. Tian, Y., et al. Simultaneous nanostructure and heterojunction engineering of graphitic carbon nitride via in situ Ag-doping for enhanced photoelectrochemical activity[J]. Applied Catalysis B: Environmental, 2015, 163: 611-622.

[13] Ong W. J., Tan L. L., Ng Y. H., et al. Graphitic carbon nitride (g-C_3N_4)-based photocatalysts for artificial photosynthesis and environmental remediation: are we a step closer to achieving sustainability?[J]. Chemical reviews, 2016, 116(12): 7159-7329.

[14] Wang Y., Li Y., Bai X., et al. Facile synthesis of Y-doped graphitic carbon nitride with enhanced photocatalytic performance[J]. Catalysis Communications, 2016, 84: 179-182.

[15] Gao H., Yan S., Wang J., et al. Ion coordination significantly enhances the photocatalytic activity of graphitic-phase carbon nitride[J]. Dalton Transactions, 2014, 43(22): 8178-8183.

[16] Yuan X., Zhou C., Jin Y., et al. Facile synthesis of 3D porous thermally exfoliated g-C_3N_4 nanosheet with enhanced photocatalytic degradation of organic dye[J]. Journal of Colloid and Interface Science, 2016, 468: 211-219.

Chapter 2

CeO$_2$/g-C$_3$N$_4$ nanocomposite

2.1 Background

With the increase of environmental pollution and higher need of clean energy, the development of photocatalytic H$_2$ production technologies for splitting water has received increasing public attention [1-3]. It is reported that hydrogen energy could be obtained by photocatalytic water splitting in 1972[4]. After that, lots of photocatalytic materials were developed with the physical structures, optical and electrical properties to achieve high photocatalytic activity[5,6].

Graphitic carbon nitride abbreviated to g-C$_3$N$_4$ could be considered as the promising candidate material for photocatalytic hydrogen production because of its medium band gap with visible-light absorbance and excellent photocatalytic stability, which is firstly investigated by Wang et al.[7]. However, it is reported that g-C$_3$N$_4$ has some drawbacks with a relatively high charge carrier recombination rate, marginal visible-light absorption and electrical conductivity to limit its photocatalytic activity[8]. Therefore, g-C$_3$N$_4$ based photocatalysts are developed by synthesis methods[9]. Nanostructure design with high charge carries separation rate[10] and electronic structure modulation for faster charge transfer[11,12]. The g-C$_3$N$_4$ based composites, for example carbon/g-C$_3$N$_4$[13], Cu$_2$O/g-C$_3$N$_4$[14], CoO/g-C$_3$N$_4$[15], Co$_2$P/g-C$_3$N$_4$[16], Bi$_2$MoO$_6$/g-C$_3$N$_4$[17], are constructed to promote the photocatalytic H$_2$ evolution performance form water splitting. These g-C$_3$N$_4$ based composites with synergistic effects at the interface have superior photocatalytic performance than that of the carbon nitride, due to faster electron transfer and more solar light absorbance and efficacious photo-generated electrons and holes separation[18,19].

Ceria dioxide (CeO$_2$), as one of the common and environmentally friendly semiconductors, has the valence change between Ce^{4+} and Ce^{3+} to be widely used as a catalytic material[20]. It was reported that crystal CeO$_2$ combined with g-C$_3$N$_4$

showed an excellent photocatalyst for splitting water to generate H_2 with high visible-light absorbance[21]. Zou et al. reported physical mixed cubic CeO_2 and carbon nitride to prepare CeO_2/g-C_3N_4 composite with enhancement of photocatalytic H_2 production under visible-light illumination[22]. However, physical mixing method is hardly to achieve CeO_2 in the composite materials with high disperse and the close-knit interface[23]. The recombination of photo-generated electron and hole pairs could be retarded and charge transfer could be fast by making highly dispersed CeO_2/g-C_3N_4 material[24]. It is reported that facile thermal annealing method can be used to prepare g-C_3N_4 based composites with easy fabrication and close knit for efficient photocatalytic H_2 production from water splitting[25], but the CoO nanoparticles in the g-C_3N_4 based composites are likely to aggregate. It is well known that the nanoparticles can be highly dispersed by rotation-evaporation method. Therefore, photocatalytic hydrogen generation performance of CeO_2/g-C_3N_4 composite will be improved by facile thermal annealing and rotation-evaporation method. Besides, there is no literature on this construction of CeO_2/g-C_3N_4 material for photocatalytic hydrogen-generated from water-splitting by this facilely prepared method.

Our research team synthesized CeO_2/g-C_3N_4 composite which is first synthesized by utilizing a facile thermal annealing combined with rotary evaporation approach in a nitrogen atmosphere, targeting photocatalytic H_2 production from water splitting driven by visible light. We employed a series of characterization methods to investigate the physical structure and chemical properties of the CeO_2/g-C_3N_4 materials. Furthermore, by conducting comprehensive comparisons with pure g-C_3N_4, we explored the developed photocatalytic H_2 generation performance and elucidated the reaction mechanism of the as-synthesized CeO_2/g-C_3N_4 composite.

2.2 Preparation of CeO_2/g-C_3N_4 composite

Urea, cerium acetate and TEOA (triethanolamine) of the raw chemical materials with analytical reagent were supplied by Aladdin Industrial Corporation. The graphitic carbon nitride samples were obtained by thermal calcination of urea[26]. Firstly, a certain amount of urea was placed into a crucible with a cover under the calcination temperature of 500 °C for 3 h in the muffle furnace. The heating rate of temperature was 10 °C/min. After gradually cooling down to room temperature, 300 mg g-C$_3$N$_4$ powders were put into the deionized H$_2$O with the volume of 50 mL and were then transferred into water bath at 70 °C after sonication. Then, a certain volume of 2 mg/mL Ce(NO$_3$)$_2$ aqueous solution was gradually added into the prepared g-C$_3$N$_4$ samples and kept stirring under room temperature for 20 h to form homogeneous solution. After the mixture was rotavaporated and dried in the rotavapor, the left material was annealed with nitrogen atmosphere at 400 °C for 2 h in the tube furnace to prepare CeO$_2$/g-C$_3$N$_4$ nanocomposites. The mass ratio from 0.5% to 7% of Ce to carbon nitride in the composite was prepared by this facile thermal annealing method.

2.3   Characterization of CeO$_2$/g-C$_3$N$_4$ composite

2.3.1   XRD of CeO$_2$/g-C$_3$N$_4$ composite

The crystalline structures of as-synthesized pure g-C$_3$N$_4$ and CeO$_2$/g-C$_3$N$_4$ nanocomposite were measured by X-Ray diffraction(XRD). The CeO$_2$/g-C$_3$N$_4$ nanocomposite and fresh g-C$_3$N$_4$ show the palpable diffraction peaks at 13.0° and 27.4° in Figure 2.1, which could be ascribed to (100) and (002) reflections of carbon

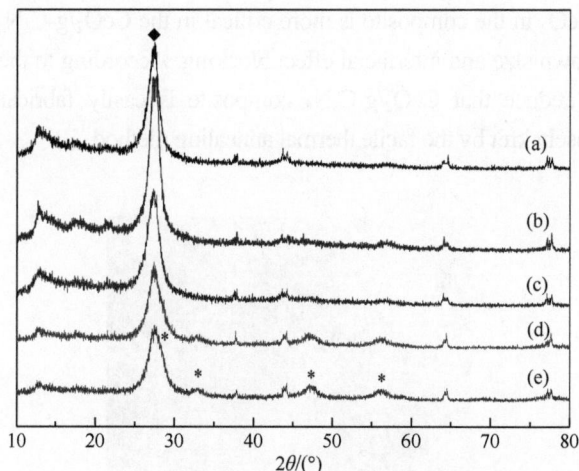

Figure 2.1   XRD patterns of (a)g-C$_3$N$_4$, (b)0.5% CeO$_2$/g-C$_3$N$_4$, (c)1% CeO$_2$/g-C$_3$N$_4$, (d) 5% CeO$_2$/g-C$_3$N$_4$ and, (e)7% CeO$_2$/g-C$_3$N$_4$ nanocomposite

nitride, respectively. It is assumed to be the flat interlayer structure packing unit in the g-C_3N[27]. Four diffraction peaks at 28.5°, 33.1°, 47.4° and 56.3° are ascribed to (111), (200), (220) and (311) crystalline structures of CeO_2 (JCPDS 34-0394), respectively[28-30]. The intensity of these diffraction peaks is gradually obvious with increase of the CeO_2 nanoparticles content, indicating that the crystalline structures of CeO_2 are improved in CeO_2/g-C_3N_4 composite with more Ce content.

2.3.2 TEM of CeO₂/g-C₃N₄ composite

The morphologies of the synthesized materials were characterized by transmission electron microscopy (TEM), and the TEM images are presented in Figure 2.2. From the results of Figure 2.2(c) and Figure 2.2(e), highly dispersed CeO_2 nanoparticles are observed onto fresh g-C_3N_4 sheets. The HRTEM pictures of CeO_2/g-C_3N_4 composite are shown in the Figure 2.2(d) and Figure 2.2(f). From the enlarged high resolution TEM images [Figure 2.2(f)] of 5% CeO_2/g-C_3N_4, the 0.31 nm and 0.27 nm lattice fringes spacing could be observed, which is ascribed to (111) and (200) crystalline structures of exposed CeO_2 in the composite, respectively[31]. It also can be seen that the (200) crystal planes of CeO_2 with 0.27 nm lattice fringes spacing of in composite with 1% Ce content. In these composites, the CeO_2 nanoparticles are successfully in-situ growing onto the g-C_3N_4 sheets with the facile thermal annealing method. The size of the CeO_2 nanoparticle is about 10 nm, which is much smaller than the size of the reported CeO_2/g-C_3N_4 composite (20 nm) prepared by physical mixing method[22]. With the increased loading amount of cerium, aggregation of CeO_2 in the composite is more critical in the CeO_2/g-C_3N_4 nanocomposite, leading to the grown size and interfacial effect blocking. According to the XRD and TEM results, we can deduce that CeO_2/g-C_3N_4 composite is easily fabricated to be highly dispersed and closely knit by the facile thermal annealing method.

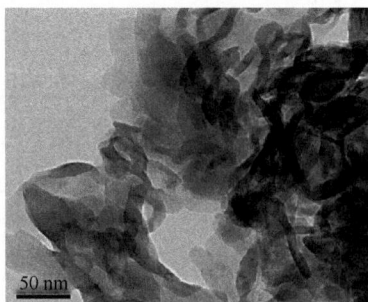

(a) TEM images of g-C₃N₄

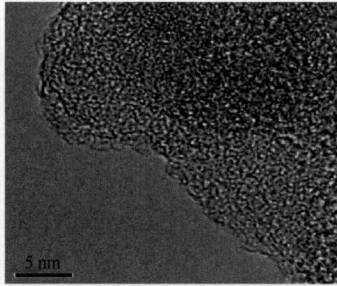

(b) high Resolution Transmission electron microscopy
(HR-TEM) images of g-C$_3$N$_4$

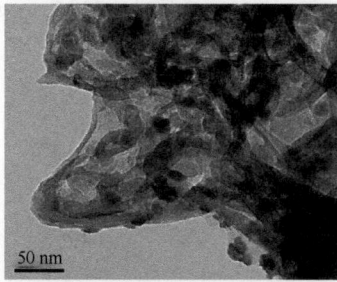

(c) TEM images of 1% CeO$_2$/g-C$_3$N$_4$ nanocomposite

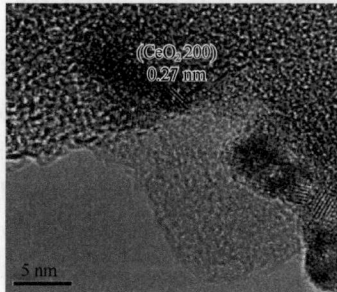

(d) HR-TEM images of 1% CeO$_2$/g-C$_3$N$_4$ nanocomposite

(e) TEM images of 5% CeO$_2$/g-C$_3$N$_4$ nanocomposite

Figure 2.2

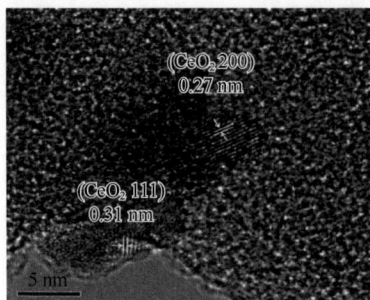

(f) HR-TEM images of 5% $CeO_2/g-C_3N_4$ nanocomposite

Figure 2.2   TEM and HR-TEM images of $g-C_3N_4$, 1% $CeO_2/g-C_3N_4$ nanocomposite, 5% $CeO_2/g-C_3N_4$ nanocomposite

2.3.3   XPS of $CeO_2/g-C_3N_4$ composite

Surface chemical compositions of as-prepared fresh $g-C_3N_4$ and $CeO_2/g-C_3N_4$ nanocomposite are further studied by X-ray photoelectron spectroscopy(XPS), and the obtained results are shown in the Figure 2.3(a)-(e). In Figure 2.3(a), there are three sharp peaks at 530.2 eV, 398.9 eV and 285.1 eV in the prepared graphitic carbon nitride and composites, which can be attributed to the O 1s, N 1s and C 1s signals, respectively. The strong signals of C 1s and N 1s are observed, indicating that in the prepared samples carbon and nitrogen elements are the main component. It can be seen that there are three obvious peaks at 284.8 eV, 288.1 eV, and 293.6 eV of C 1s spectra in Figure 2.3(b) ,which could correspond to the signals of carbon contamination, the sp^2-bonded carbon in N=C—N groups of carbon nitride and π ring, respectively[24]. N 1s spectra in Figure 2.3(c) could be divided into four peaks. The 398.7 eV and 399.4 eV peaks are corresponding to sp^2-bonded nitrogen atoms in C=N—C of $g-C_3N_4$[32] and tertiary nitrogen in N—C_3 groups[33], respectively. The 400.9 eV and 404.7 eV peaks are according to the N—H_2 groups[32] and positive charge located in heterocycles[34], respectively. The Figure 2.3(d) and Figure 2.3(e) present the O 1s spectra and Ce 3d spectra, respectively. Form the results of the O 1s spectra in Figure 2.3(d), it can be found that there are two peaks in the O 1s spectra at 528.9 eV and 532.2 eV, which are assumed to the lattice and adsorbed oxygen in CeO_2, respectively[22]. It is found that the signal of lattice oxygen becomes stronger and the peak at about 532.2 eV of adsorbed oxygen in CeO_2

shows a broader feature with the increase of Ce loading. It is reported that the broad peak changes by the electronic state of the O species in the adsorbed water molecule and hydroxyl groups[35]. As the electronegativity N atom is weaker than O atom, the electron will prefer to transfer onto O atom by a typical hydrogen bond. Therefore, the binding energy of the O species in the adsorbed hydroxyl groups becomes smaller. The interfacial effect between ceria dioxide and carbon nitride could be enhanced to improve the catalytic H_2 evolution performance due to adsorbed OH groups on the surface of CeO₂/g-C₃N₄ nanocomposite. Figure 2.3(e) shows the Ce 3d spectra of 1% and 5% CeO₂/g-C₃N₄. The Ce 3d peaks of 1% CeO₂/g-C₃N₄ at 903.9 eV and 885.2 eV could verify existence of the ceria element. There are eight peaks divided from Ce 3d XPS spectra of the prepared composites. The signals of the peaks at 900.5 eV(u), 881.8 eV(v), 916.1 eV (u‴) and 898.0 eV (v‴) are weaker than those of the peaks at 911.0 eV (u″) and 891.0 eV (v″), suggesting the presence of Ce^{4+}. Ce^{4+} could be the dominant state in the composite because of the high intensity and area of u‴ (916.1 eV) corresponding to the signal of Ce^{4+}. The peaks at 903.8 eV (u), 885.1 eV (v), 900.5 eV (u) and 881.8 eV (v) can be corresponding to the signal of Ce^{3+}[24]. It is reported that Ce^{3+} is an indicator for the photo-reduction reaction[36], which will develop the catalytic H_2 production performance from water-splitting.

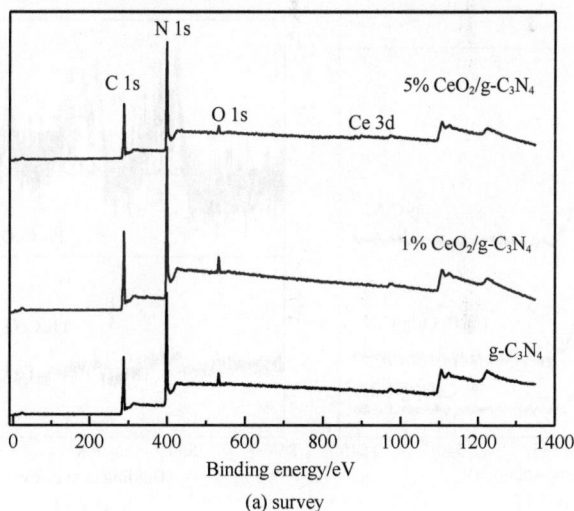

(a) survey

Figure 2.3

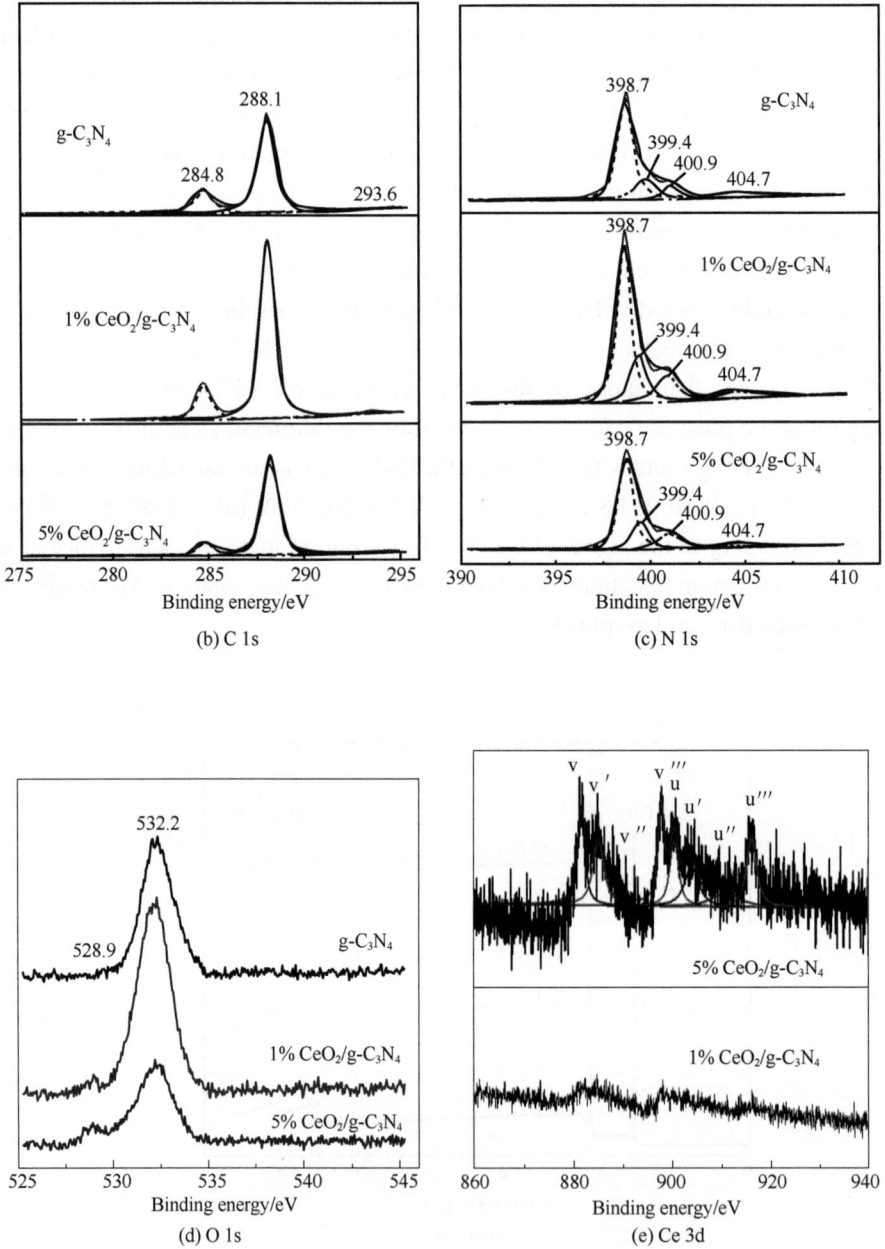

(b) C 1s

(c) N 1s

(d) O 1s

(e) Ce 3d

Figure 2.3   XPS profiles of the prepared samples

2.3.4   UV-vis absorption and PL spectra of CeO_2/g-C_3N_4 composite

Generally, the optical and photoelectric properties are of cardinal significance for catalytic hydrogen production activity from water splitting. UV-Vis absorption spectra are measured to identify the optical properties of synthesized fresh g-C_3N_4 and CeO_2/g-C_3N_4 nanocomposite. Figure 2.4 presents UV-vis absorption spectra of carbon nitride and the composites with different CeO_2 contents. It is found that 1% CeO_2/g-C_3N_4 exhibits the best ultraviolet and visible-light absorbance. The enhanced visible-light absorption suggests that the CeO_2/g-C_3N_4 nanocomposite can use more visible light, which will enhance the photocatalytic activity for H_2 production. To identify the photoelectric properties of as-synthesized fresh g-C_3N_4 and CeO_2/g-C_3N_4 composite, photoluminescence (PL) is measured and the results are shown in Figure 2.5. PL spectra of obtained fresh g-C_3N_4 and the nanocomposite are measured with the excitation wavelength at 370 nm. It is found that the signal of the prepared composite emission profile becomes weaker with an increase in the loading content of CeO_2, suggesting the faster photo-generated electron transfer on the surface of the composite[37,38]. Therefore, the recombination rate of photo-generated charge carriers could be significantly restrained by construction of the CeO_2/g-C_3N_4 composite with the facile thermal annealing method.

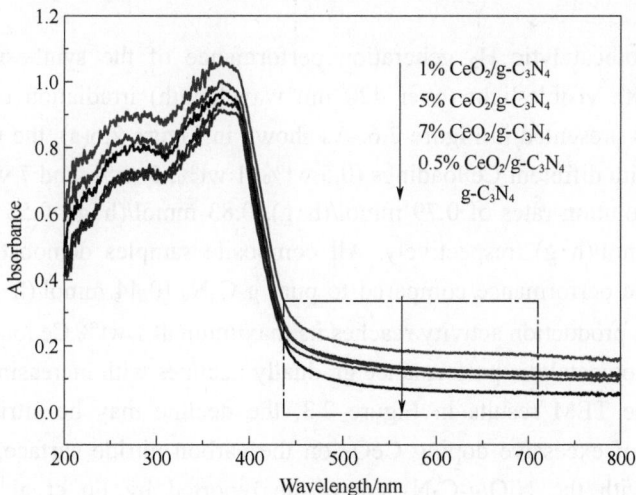

Figure 2.4   UV-vis absorption spectra of the synthesized g-C_3N_4, 0.5% CeO_2/g-C_3N_4, 1% CeO_2/g-C_3N_4, 5% CeO_2/g-C_3N_4 and 7% CeO_2/g-C_3N_4 nanocomposite

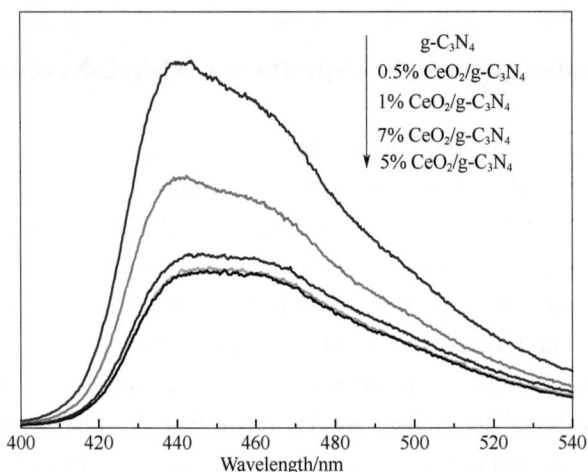

Figure 2.5   PL spectra (λ_{ex} = 370 nm) for the synthesized g-C_3N_4, 0.5% CeO_2/g-C_3N_4, 1% CeO_2/g-C_3N_4, 5% CeO_2/g-C_3N_4 and 7% CeO_2/g-C_3N_4 nanocomposite

2.4   Photocatalytic H$_2$ generation testing of CeO$_2$/g-C$_3$N$_4$ composite

2.4.1   Hydrogen production efficiency

The photocatalytic H_2 generation performance of the synthesized samples measured with visible-light (over 420 nm wavelength) irradiation using Pt as a co-catalyst is presented in Figure 2.6. As shown in Figure 2.6(a), the CeO_2/g-C_3N_4 composite with different Ce loadings (0.5 wt%, 1 wt%, 5 wt%, and 7 wt%) exhibits hydrogen evolution rates of 0.79 mmol/(h·g), 0.83 mmol/(h·g), 0.55 mmol/(h·g), and 0.53 mmol/(h·g), respectively. All composite samples demonstrate superior photocatalytic performance compared to pure g-C_3N_4 [0.44 mmol/(h·g)]. Notably, the hydrogen production activity reaches its maximum at 1 wt% Ce loading, beyond which the photocatalytic performance gradually declines with increasing Ce content. Based on the TEM results in Figure 2.3, the decline may be attributed to the aggregation of excessive doping CeO_2 on the carbon nitride surface, which is in agreement with the NiO/g-C_3N_4 composite reported by Fu et al.[39]. The 1% CeO_2/g-C_3N_4 nanocomposite has superior photocatalytic performance with the

16

average H$_2$ generation rate of 0.83 mmol/ (h·g), about 100% higher than pure g-C$_3$N$_4$. Based on comparison with previous reports (shown in Table 2.1), this hydrogen evolution rate [0.83 mmol/(h·g)] is much more effective than that of most reported carbon nitride based photocatalytic materials, such as Cu$_2$O[14], CNT[40], CoO[25], GO[41], Bi$_2$MoO$_6$[17] and so on, revealing the excellent synergetic effect between CeO$_2$ and carbon nitride. More importantly, the amount of hydrogen evolution of the CeO$_2$/g-C$_3$N$_4$ nanocomposite prepared by the facile thermal annealing method with 1.5w% Pt is almost the same as the CeO$_2$/g-C$_3$N$_4$ composite prepared by the physical mixing method with 3w% Pt reported by Zou et al.[21,22].

The photocatalytic stability of hydrogen generation of 1% CeO$_2$/g-C$_3$N$_4$ nanocomposite is measured by four cycling experiments, and the results are shown in Figure 2.6(b). After four recycling runs with visible-light (over 420 nm wavelength) illumination, the photocatalytic H$_2$ evolution performance of 1% CeO$_2$/g-C$_3$N$_4$ exhibits favourable stability. Based on these results, it can be safely concluded that CeO$_2$/g-C$_3$N$_4$ nanocomposite prepared by the facile thermal annealing method exhibits outstanding H$_2$ generation performance and stability due to the enhancement of visible-light absorbance and effective separation of electron-hole pairs.

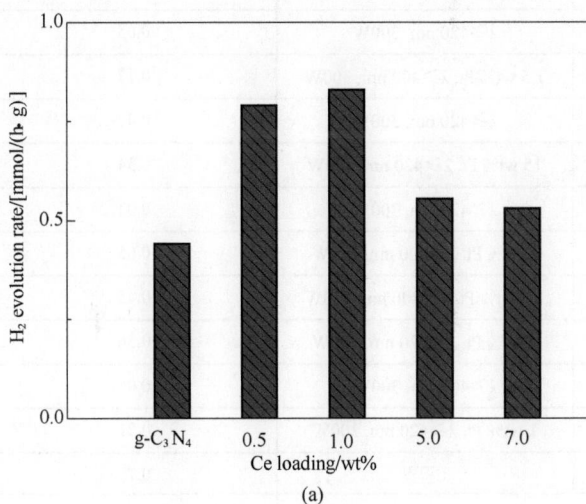

(a)

Figure 2.6

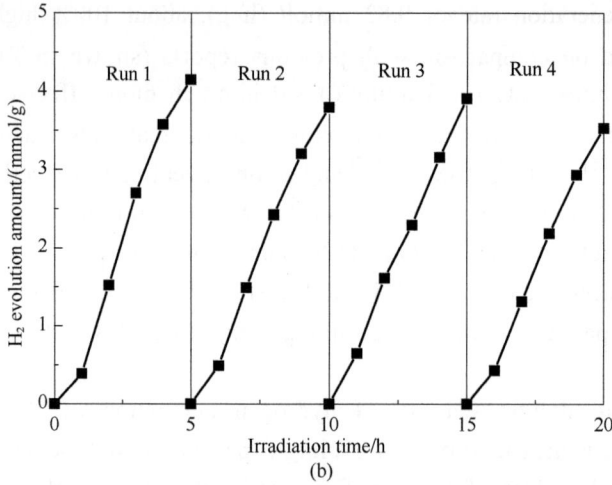

(b)

Figure 2.6   (a) The rate of photocatalytic H_2 evolution with 10 vol% TEOA, 1.5 wt% Pt, and 50 mg 1 wt% CeO_2/g-C_3N_4 photocatalysts under visible light ($\lambda \geqslant 420$ nm), (b) recyclability of 1 wt% CeO_2/g-C_3N_4 photocatalyst for the photocatalytic H_2 evolution under visible-light irradiation ($\lambda \geqslant 420$ nm)

Table 2.1   Summary of the photocatalytic H_2 evolution rate on g-C_3N_4-based photocatalysts

Photocatalysts	Reaction conditions	H_2 evolution rate/[mmol/(h·g)]	Reference
CeO_2/g-C_3N_4	1.5 wt% Pt, $\lambda \geqslant 420$ nm, 300W	0.83	In this book
Cu_2O/g-C_3N_4	3 wt% Pt, $\lambda \geqslant 420$ nm, 300W	0.24	[14]
Co_2P /g-C_3N_4	$\lambda \geqslant 420$ nm, 300W	0.05	[16]
CNT/g-C_3N_4	1.5 wt% Pt, $\lambda \geqslant 400$ nm, 300W	0.17	[40]
Ni_2P/g-C_3N_4	$\lambda \geqslant 420$ nm, 300W	0.47	[42]
Ni_2P/g-C_3N_4	15 wt% Pt, $\lambda \geqslant 420$ nm, 300W	3.34	[43]
NiO/g-C_3N_4	$\lambda \geqslant 420$ nm, 300W	0.03	[39]
CoO/g-C_3N_4	3 wt% Pt, $\lambda \geqslant 400$ nm, 300W	0.65	[25]
GO/g-C_3N_4	1.5 wt% Pt, $\lambda \geqslant 400$ nm, 350W	0.45	[41]
Bi_2MoO_6/g-C_3N_4	3 wt% Pt, $\lambda \geqslant 420$ nm, 300W	0.56	[17]
CoO/g-C_3N_4	$\lambda \geqslant 400$ nm, 300W	0.05	[15]
Carbon/g-C_3N_4	1 wt% Pt, $\lambda \geqslant 420$ nm, 300W	0.21	[13]
TiO_2/g-C_3N_4	150W	0.77	[12]
Cr_2O_3/g-C_3N_4	5 wt% Pt, $\lambda \geqslant 420$ nm, 300W	0.21	[44]
g-C_3N_4(580)-T	3 wt% Pt, $\lambda \geqslant 420$ nm, 300W	1.39	[45]

continued

Photocatalysts	Reaction conditions	H$_2$ evolution rate/[mmol/(h·g)]	Reference
PtNi/g-C$_3$N$_4$	λ⩾420 nm, 300W	2.0	[46]
g-C$_3$N$_4$/Nb$_2$O$_5$	1.5 wt% Pt, λ⩾420 nm, 300W	1.71	[47]
Cu/g-C$_3$N$_4$	λ⩾420 nm, 300W	0.47	[48]
Fe$_2$O$_3$/g-C$_3$N$_4$	5 vol% Pt, λ⩾400 nm, 300W	0.78	[49]
SrTiO$_3$/g-C$_3$N$_4$	1 wt% Pt, λ⩾420 nm, 300W	0.44	[50]
CDs/g-C$_3$N$_4$	3 wt% Pt, λ⩾420 nm, 300W	2.34	[51]
CuS/g-C$_3$N$_4$	λ⩾420 nm, 300W	0.017	[52]
FeCoP/g-C$_3$N$_4$	λ⩾420 nm, 300W	0.35	[53]
Cubic CeO$_2$/g-C$_3$N$_4$	3 wt% Pt, λ⩾420 nm, 300W	0.86	[21]
Cubic CeO$_2$/g-C$_3$N$_4$	3 wt% Pt, λ⩾420 nm, 300W	0.86	[22]

2.4.2   Charge separation and transfer performance

For the sake of examining the mechanism for development of photocatalytic capacity of CeO$_2$/g-C$_3$N$_4$ nanocomposite, EIS and photocurrents were carried out to obtain charge separation and transfer properties. The transient PC responses of prepared fresh g-C$_3$N$_4$ and 1% CeO$_2$/g-C$_3$N$_4$ nanocomposite were measured under visible-light illumination with the interval light on/off cycles, and the results are shown in Figure 2.7(a). It is observed that as-prepared 1% CeO$_2$/g-C$_3$N$_4$ composite exhibits excellent photocatalytic stability test and obtains bigger photocurrent than fresh g-C$_3$N$_4$ sample in Figure 2.7(a), suggesting that the interfacial charge transfer as well as electron-hole separation process between CeO$_2$ and g-C$_3$N$_4$ is largely improved. Figure 2.7(b) shows the Nyquist impedance plots of the as-prepared carbon nitride and 1% CeO$_2$/g-C$_3$N$_4$ composite under darkness. The surface reaction rate of the electrode can be determined by the arc radius. The smaller arc radius means the higher effective electron-hole pairs separation and the faster electron transfer across the electrode/electrolyte[15]. In Figure 2.7(b), it can be clearly observed that the arc radius of 1% CeO$_2$/g-C$_3$N$_4$ electrode is smaller than the prepared carbon nitride electrode, indicating that prepared CeO$_2$/g-C$_3$N$_4$ nanocomposite exhibits the faster electron transfer with effective electron-hole pairs separation. The photoelectric properties analysis are in agreement with the results of PL spectra, suggesting that the enhancement of activity for H$_2$ evolution of CeO$_2$/g-C$_3$N$_4$ composite is majorly ascribed to faster interfacial electron transfer and the higher effective electron-hole pairs separation.

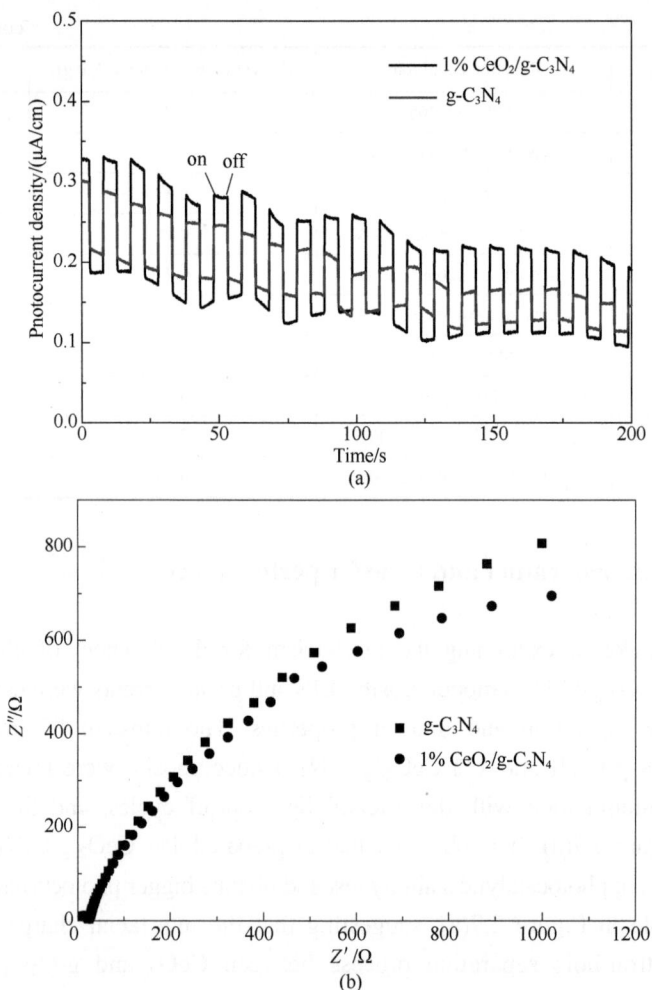

Figure 2.7 (a) Transient photocurrent density of g-C$_3$N$_4$ and 1% CeO$_2$/g-C$_3$N$_4$ electrodes at 0.3 V versus Ag/AgCl, (b) electrochemical impedance spectra of the as-prepared samples at -0.4 V versus Ag/AgCl

2.4.3 Photocatalytic H$_2$ evolution mechanism

Based on these above analysis and experimental results, the conceivable mechanism of hydrogen generation of CeO$_2$/g-C$_3$N$_4$ nanocomposite is proposed and the schematic illustrations are shown in Figure 2.8. Firstly, the electron-hole pairs are produced by carbon nitride with enough energy from the photons of visible light illumination. Secondly, the charge in the conduction band of g-C$_3$N$_4$ will devolve onto the surface of CeO$_2$ and platinum particles to catalyze the reduction of protons to H$_2$, while in the valence band of carbon nitride, the left holes will be quenched by sacrificial reagents (TEOA). From the results of Ce 3d XPS spectra in Figure 2.3(e), both Ce^{4+} (the peak at 916.1 eV) and Ce^{3+} (the peak at 903.8 eV) exist in the CeO$_2$/g-C$_3$N$_4$ nanocomposite. The generated electrons can be trapped by Ce^{4+} to partially reduce Ce^{4+} to Ce^{3+} at the interface of the CeO$_2$/g-C$_3$N$_4$ nanocomposite[54,55]. Ce^{3+} could react with protons to produce H$_2$ with platinum particles. Thus, the electron-hole recombination is greatly retarded and the charge transfer is more effective, enabling the development of H$_2$ evolution from the H$_2$O reduction of the CeO$_2$/g-C$_3$N$_4$ nanocomposite.

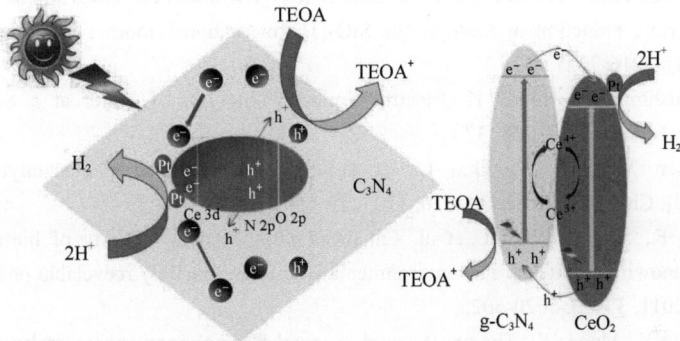

Figure 2.8   Schematic illustrations of the proposed photocatalytic H$_2$ evolution mechanism over the CeO$_2$/g-C$_3$N$_4$ nanocomposite under visible light irradiation

2.5 Conclusion

The novel CeO$_2$/g-C$_3$N$_4$ nanocomposite with different contents of CeO$_2$ particles was facilely prepared to evaluate the photocatalytic H$_2$ production performance. Based on the experimental results, the CeO$_2$/g-C$_3$N$_4$ composite with 1wt% Ce loading exhibits the best activity,with an H$_2$ generation rate of 0.83 mmol/(h·g) under   visible-light illumination, which is around 100% higher than that of fresh carbon nitride. The remarkably improved capacity for hydrogen evolution of the CeO_2/g-C_3N_4 composite is mainly ascribed to faster interfacial charge transfer and the more effective photogenerated electronhole pairs separation. This work reveals that the interfacial effect is a critical factor for the development of photocatalytic H_2 evolution activity.

Reference

[1] An Z., Gao J., Wang L., et al. Novel microreactors of polyacrylamide (PAM)CdS microgels for admirable photocatalytic H_2 production under visible light[J]. International Journal of Hydrogen Energy, 2019, 44 (3):1514-1524.

[2] Fajrina N., Tahir, M. A critical review in strategies to improve photocatalytic water splitting towards hydrogen production[J]. International Journal of Hydrogen Energy, 2019, 44 (2): 540-577.

[3] Sun H., Chen J., Liu S., et al. Photocatalytic H_2 evolution of porous silicon derived from magnesiothermic reduction of mesoporous SiO_2[J]. International Journal of Hydrogen Energy, 2019, 44 (14): 7216-7221.

[4] Fujishima A., Honda K. Electrochemical Photolysis of Water at a Semiconductor Electrode[J]. Nature, 1972, 238: 37.

[5] Chen X., Shen S., Guo L., et al.Semiconductor-based Photocatalytic Hydrogen Generation[J]. Chemical Reviews, 2010, 110 (11): 6503-6570.

[6] Xu F., Shen Y., Sun L.,et al. Enhanced photocatalytic activity of hierarchical ZnO nanoplate-nanowire architecture as environmentally safe and facilely recyclable photocatalyst[J]. Nanoscale, 2011, 3 (12): 5020-5025.

[7] Wang X., Maeda K., Thomas A., et al. A metal-free polymeric photocatalyst for hydrogen production from water under visible light[J]. Nature Materials, 2008, 8: 76.

[8] Zhao S., Zhang Y., Zhou Y., et al. Facile one-step synthesis of hollow mesoporous g-C_3N_4 spheres with ultrathin nanosheets for photoredox water splitting[J]. Carbon, 2018, 126: 247-256.

[9] Xu Q., Cheng B., Yu J., et al. Making co-condensed amorphous carbon/g-C_3N_4 composites with improved visible-light photocatalytic H_2-production performance using Pt as cocatalyst[J]. Carbon, 2017, 118: 241-249.

[10] Martin D. J., Qiu K., Shevlin S. A., et al. Highly Efficient Photocatalytic H_2 Evolution from Water using Visible Light and Structure-Controlled Graphitic Carbon Nitride[J]. Angewandte Chemie International Edition, 2014, 53 (35): 9240-9245.

[11] Zhang Y., Liu J., Wu G.,et al.Porous graphitic carbon nitride synthesized via direct polymerization of urea for efficient sunlight-driven photocatalytic hydrogen production[J]. Nanoscale, 2012, 4 (17): 5300-5303.

[12] Fettkenhauer C., Clavel G., Kailasam K., et al. Facile synthesis of new, highly efficient

SnO$_2$/carbon nitride composite photocatalysts for the hydrogen evolution reaction[J]. Green Chemistry, 2015, 17 (6): 3350-3361.

[13] Yan J., Wu H., Chen H., et al. Fabrication of TiO$_2$/C$_3$N$_4$ heterostructure for enhanced photocatalytic Z-scheme overall water splitting[J]. Applied Catalysis B: Environmental, 2016, 191: 130-137.

[14] Chen J., Shen S., Guo P., et al. In-situ reduction synthesis of nano-sized Cu$_2$O particles modifying g-C$_3$N$_4$ for enhanced photocatalytic hydrogen production[J]. Applied Catalysis B: Environmental, 2014, 152-153: 335-341.

[15] Guo F., Shi W., Zhu C., et al. CoO and g-C$_3$N$_4$ complement each other for highly efficient overall water splitting under visible light[J]. Applied Catalysis B: Environmental, 2018, 226: 412-420.

[16] Zeng D., Ong W. J., Chen Y., et al. Co$_2$P Nanorods as an Efficient Cocatalyst Decorated Porous g-C$_3$N$_4$ Nanosheets for Photocatalytic Hydrogen Production under Visible Light Irradiation[J]. Particle & Particle Systems Characterization, 2018, 35, (1): 1700251.

[17] Li J., Yin Y., Liu E.,et al. In situ growing Bi$_2$MoO$_6$ on g-C$_3$N$_4$ nanosheets with enhanced photocatalytic hydrogen evolution and disinfection of bacteria under visible light irradiation[J]. Journal of Hazardous Materials, 2017, 321: 183-192.

[18] Zhao Z., Sun Y., Dong F. Graphitic carbon nitride based nanocomposites: a review[J]. Nanoscale, 2015, 7 (1): 15-37.

[19] Fu, J., Xu, Q., Low, J., et al. Ultrathin 2D/2D WO$_3$/g-C$_3$N$_4$ step-scheme H$_2$-production photocatalyst. Applied Catalysis B: Environmental, 2019, 243: 556-565.

[20] Wang, Y., Wang, F., Chen, Y., et al.,Enhanced photocatalytic performance of ordered mesoporous Fe-doped CeO$_2$ catalysts for the reduction of CO$_2$ with H$_2$O under simulated solar irradiation. Applied Catalysis B: Environmental 2014, 147: 602-609.

[21] Zou, W., Deng, B., Hu, X., et al. Crystal-plane-dependent metal oxide-support interaction in CeO$_2$/g-C$_3$N$_4$ for photocatalytic hydrogen evolution. Applied Catalysis B: Environmental, 2018, 238: 111-118.

[22] Zou W., Shao Y., Pu Y., et al. Enhanced visible light photocatalytic hydrogen evolution via cubic CeO$_2$ hybridized g-C$_3$N$_4$ composite[J]. Applied Catalysis B: Environmental, 2017, 218: 51-59.

[23] Yu Y., Zhong Q., Cai W., et al. Promotional effect of N-doped CeO$_2$ supported CoO$_x$ catalysts with enhanced catalytic activity on NO oxidation[J]. Journal of Molecular Catalysis A: Chemical, 2015, 398: 344-352.

[24] Li M., Zhang L., Wu M., et al. Mesostructured CeO$_2$/g-C$_3$N$_4$ nanocomposites: Remarkably enhanced photocatalytic activity for CO$_2$ reduction by mutual component activations[J]. Nano Energy, 2016, 19: 145-155.

[25] Mao Z., Chen J., Yang Y., et al. Novel g-C$_3$N$_4$/CoO Nanocomposites with Significantly Enhanced Visible-Light Photocatalytic Activity for H$_2$ Evolution[J]. ACS Applied Materials & Interfaces, 2017, 9 (14): 12427-12435.

[26] Dong F., Wu L., Sun Y., et al. Efficient synthesis of polymeric g-C$_3$N$_4$ layered materials as novel efficient visible light driven photocatalysts[J]. Journal of Materials Chemistry, 2011, 21

(39): 15171-15174.

[27] Yang X., Chen Z., Xu J., et al. Tuning the Morphology of g-C$_3$N$_4$ for Improvement of Z-Scheme Photocatalytic Water Oxidation[J]. ACS Applied Materials & Interfaces, 2015, 7 (28): 15285-15293.

[28] Magdalane C. M., Kaviyarasu K., Vijaya J. J., et al. Evaluation on the heterostructured CeO$_2$/Y$_2$O$_3$ binary metal oxide nanocomposites for UV/Vis light induced photocatalytic degradation of Rhodamine-B dye for textile engineering application[J]. Journal of Alloys and Compounds, 2017, 727: 1324-1337.

[29] Magdalane C. M., Kaviyarasu K., Vijaya J. J., et al. Photocatalytic activity of binary metal oxide nanocomposites of CeO$_2$/CdO nanospheres: Investigation of optical and antimicrobial activity[J]. Journal of Photochemistry and Photobiology B: Biology, 2016, 163: 77-86.

[30] Maria Magdalane C., Kaviyarasu K., Judith Vijaya J., et al. Photocatalytic degradation effect of malachite green and catalytic hydrogenation by UV-illuminated CeO$_2$/CdO multilayered nanoplatelet arrays: Investigation of antifungal and antimicrobial activities[J]. Journal of Photochemistry and Photobiology B: Biology, 2017, 169: 110-123.

[31] Maria Magdalane C., Kaviyarasu K., Judith Vijaya J., et al. Facile synthesis of heterostructured cerium oxide/yttrium oxide nanocomposite in UV light induced photocatalytic degradation and catalytic reduction: Synergistic effect of antimicrobial studies[J]. Journal of Photochemistry and Photobiology B: Biology, 2017, 173: 23-34.

[32] Gao D., Xu Q., Zhang J., et al. Defect-related ferromagnetism in ultrathin metal-free g-C$_3$N$_4$ nanosheets[J]. Nanoscale, 2014, 6 (5): 2577-2581.

[33] Liu C., Jing L., He L.,et al. Phosphate-modified graphitic C$_3$N$_4$ as efficient photocatalyst for degrading colorless pollutants by promoting O$_2$ adsorption[J]. Chemical Communications, 2014, 50 (16): 1999-2001.

[34] Wang S., Li C., Wang T., et al. Controllable synthesis of nanotube-type graphitic C$_3$N$_4$ and their visible-light photocatalytic and fluorescent properties[J]. Journal of Materials Chemistry A, 2014, 2 (9): 2885-2890.

[35] Carrasco J., López-Durán D., Liu Z., et al. In Situ and Theoretical Studies for the Dissociation of Water on an Active Ni/CeO$_2$ Catalyst: Importance of Strong Metal-Support Interactions for the Cleavage of O-H Bonds[J]. Angewandte Chemie International Edition, 2015, 54 (13): 3917-3921.

[36] Li S., Zhu H., Qin Z., et al. Morphologic effects of nano CeO$_2$-TiO$_2$ on the performance of Au/CeO$_2$-TiO$_2$ catalysts in low-temperature CO oxidation[J]. Applied Catalysis B: Environmental, 2014, 144: 498-506.

[37] Kaviyarasu K., Fuku X., Mola G. T., et al. Photoluminescence of well-aligned ZnO doped CeO$_2$ nanoplatelets by a solvothermal route[J]. Materials Letters, 2016, 183: 351-354.

[38] Kaviyarasu K., Manikandan E., Nuru Z. Y., et al. Investigation on the structural properties of CeO$_2$ nanofibers via CTAB surfactant[J]. Materials Letters, 2015, 160: 61-63.

[39] Fu Y., Liu C. a., Zhu C.,et al. High-performance NiO/g-C$_3$N$_4$ composites for visible-light-driven photocatalytic overall water splitting[J]. Inorganic Chemistry Frontiers, 2018, 5 (7): 1646-1652.

[40] Christoforidis K. C., Syrgiannis Z., La Parola V., et al. Metal-free dual-phase full organic carbon nanotubes/g-C$_3$N$_4$ heteroarchitectures for photocatalytic hydrogen production[J]. Nano Energy, 2018, 50: 468-478.

[41] Xiang Q., Yu J., Jaroniec M. Preparation and Enhanced Visible-Light Photocatalytic H$_2$-Production Activity of Graphene/C$_3$N$_4$ Composites[J]. The Journal of Physical Chemistry C, 2011, 115 (15): 7355-7363.

[42] Zeng D., Xu W., Ong W. J., et al. Toward noble-metal-free visible-light-driven photocatalytic hydrogen evolution: Monodisperse sub-15nm Ni$_2$P nanoparticles anchored on porous g-C$_3$N$_4$ nanosheets to engineer 0D-2D heterojunction interfaces[J]. Applied Catalysis B: Environmental, 2018, 221: 47-55.

[43] Liu E., Jin C., Xu C., et al. Facile strategy to fabricate Ni$_2$P/g-C$_3$N$_4$ heterojunction with excellent photocatalytic hydrogen evolution activity[J]. International Journal of Hydrogen Energy, 2018, 43 (46): 21355-21364.

[44] Shi J., Cheng C., Hu Y., et al. One-pot preparation of porous Cr$_2$O$_3$/g-C$_3$N$_4$ composites towards enhanced photocatalytic H$_2$ evolution under visible-light irradiation[J]. International Journal of Hydrogen Energy, 2017, 42 (7): 4651-4659.

[45] Hong Y., Liu E., Shi J., et al. A direct one-step synthesis of ultrathin g-C$_3$N$_4$ nanosheets from thiourea for boosting solar photocatalytic H$_2$ evolution[J]. International Journal of Hydrogen Energy, 2019, 44 (14): 7194-7204.

[46] Peng W., Zhang S. S., Shao Y. B., et al. Bimetallic PtNi/g-C$_3$N$_4$ nanotubes with enhanced photocatalytic activity for H$_2$ evolution under visible light irradiation[J]. International Journal of Hydrogen Energy, 2018, 43 (49): 22215-22225.

[47] Huang Q. Z., Wang J. C., Wang P. P., et al. In-situ growth of mesoporous Nb$_2$O$_5$ microspheres on g-C$_3$N$_4$ nanosheets for enhanced photocatalytic H$_2$ evolution under visible light irradiation[J]. International Journal of Hydrogen Energy, 2017, 42 (10): 6683-6694.

[48] Zhang P., Song T., Wang T., et al. Effectively extending visible light absorption with a broad spectrum sensitizer for improving the H$_2$ evolution of in-situ Cu/g-C$_3$N$_4$ nanocomponents[J]. International Journal of Hydrogen Energy, 2017, 42 (21): 14511-14521.

[49] Li Y., Li F., Wang X., et al. Z-scheme electronic transfer of quantum-sized α-Fe$_2$O$_3$ modified g-C$_3$N$_4$ hybrids for enhanced photocatalytic hydrogen production[J]. International Journal of Hydrogen Energy, 2017, 42 (47): 28327-28336.

[50] Xu X., Liu G., Randorn C., et al. g-C$_3$N$_4$ coated SrTiO$_3$ as an efficient photocatalyst for H$_2$ production in aqueous solution under visible light irradiation[J]. International Journal of Hydrogen Energy, 2011, 36 (21): 13501-13507.

[51] Wang K., Wang X., Pan H., et al. In situ fabrication of CDs/g-C$_3$N$_4$ hybrids with enhanced interface connection via calcination of the precursors for photocatalytic H$_2$ evolution[J]. International Journal of Hydrogen Energy, 2018, 43 (1): 91-99.

[52] Chen T., Song C., Fan M., et al. In-situ fabrication of CuS/g-C$_3$N$_4$ nanocomposites with enhanced photocatalytic H$_2$-production activity via photoinduced interfacial charge transfer[J]. International Journal of Hydrogen Energy, 2017, 42 (17): 12210-12219.

[53] Cheng L., Xie S., Zou Y., et al. Noble-metal-free Fe$_2$P-Co$_2$P co-catalyst boosting

visible-light-driven photocatalytic hydrogen production over graphitic carbon nitride: The synergistic effects between the metal phosphides[J]. International Journal of Hydrogen Energy, 2019, 44 (8): 4133-4142.

[54] Maria Magdalane C., Kaviyarasu K., Raja A., et al. Photocatalytic decomposition effect of erbium doped cerium oxide nanostructures driven by visible light irradiation: Investigation of cytotoxicity, antibacterial growth inhibition using catalyst[J]. Journal of Photochemistry and Photobiology B: Biology, 2018, 185: 275-282.

[55] Maria Magdalane C., Kaviyarasu K., Matinise N., et al. Evaluation on La_2O_3 garlanded ceria heterostructured binary metal oxide nanoplates for UV/visible light induced removal of organic dye from urban wastewater[J]. South African Journal of Chemical Engineering, 2018, 26: 49-60.

Chapter 3

In-situ growing of CoO nanoparticles on g-C$_3$N$_4$ composite

3.1 Background

The photocatalytic activity of materials can be improved with the enhanced band structures, light absorbance, charge transport and photogenerated electron-hole pairs separation[1,2]. Several strategies, such as synthesis techniques[3], nanostructure design[4], electronic structure modulation[5,6], have been conducted to obtain highly efficient g-C$_3$N$_4$ based photocatalysts. Apart from the above mentioned methods, the Z-scheme photocatalytic system with two different photocatalysts which sparked interest from the natural photosynthetic systems of plant leaves is one of the most promising approaches to obtain hydrogen evolution from water splitting. The construction of Z-scheme g-C$_3$N$_4$-based composite is widely investigated, e.g. NiO/g-C$_3$N$_4$[7], Cu$_2$O/g-C$_3$N$_4$[8], CeO$_2$/g-C$_3$N$_4$[9], TiO$_2$/g-C$_3$N$_4$[6], Bi$_2$MoO$_6$/g-C$_3$N$_4$[10], Cr$_2$O$_3$/g-C$_3$N$_4$[11], to promote photoactivity for water splitting. In comparison with pure g-C$_3$N$_4$, these Z-scheme g-C$_3$N$_4$ based composites have superior potential merits in promoting charge transportation, limiting photogenerated electron-hole pairs recombination, strengthening light absorbance, and lowering the redox overpotential[12]. Recently, the construction of heterojunction photocatalysts is mainly focused on how to effectively limit photogenerated electron-hole pairs recombination, and it places less emphasis on the selection of semiconductors. Factly, in order to optimize the fabrication of Z-scheme g-C$_3$N$_4$ based photocatalysts for overall water splitting, it is important to design a heterojunction photocatalyst with band energy alignments capable not only of trapping an electron to effectively separate the photogenerated charges but also of suppressing the back reaction of water formation.

Cobalt monoxide (CoO), as a traditional transition metal monoxide, has gained more attention on the application of photocatalytic water splitting. It is reported that CoO exhibits the good photocatalytic water splitting activity with an extraordinary STH of around 5%[13]. Wang reported photocatalytic decomposition of water for hydrogen evolution on a $CoO/SrTiO_3$ catalyst in 2007[14]. Besides, CoO, with efficient photo-induced electron separation, can be used as an effective co-catalyst to improve the photocatalytic water splitting activity for hydrogen evolution[15]. But poor stability of the CoO catalyst is caused by H_2O_2 poisoning, which hinders its further development[16-18]. It is still a challenge to seek a suitable structure of CoO based catalyst with high activity and stability.

It is reported that the combination of g-C_3N_4 and CoO can be improved photocatalysts for water splitting[19-21]. The particles could be well dispersed on the carrier by vacuum rotation-evaporation method[22]. In this book CoO nanoparticles are in-situ growing on the g-C_3N_4 to prepare well-dispersed CoO/g-C_3N_4 composite photocatalyst by vacuum rotation-evaporation and thermal annealing method under nitrogen atmosphere. The physical structure, chemical composition, photo electric properties and photocatalytic H_2 generation activity of CoO/g-C_3N_4 nanocomposite with different Co loading are investigated in detail by a series of characterizations. The enhancement mechanism of photocatalytic overall water splitting for hydrogen evolution of as-synthesized CoO/g-C_3N_4 nanocomposite is also discussed.

3.2　Preparation of CoO/g-C_3N_4 composite

Urea, triethanolamine, and cobalt nitrate of analytical grade are purchased from Aladdin Industrial Corporation (Shanghai, China). Firstly, g-C_3N_4 is prepared by the thermal polycondensation of urea[23]. Typically, 10 g urea is placed into a covered crucible and heated at 500 °C for 3 h using a heating rate of 10 °C /min in a muffle furnace to obtain g-C_3N_4. 200 mg g-C_3N_4 powder is dispersed in 50 mL of deionized water by sonication. According to the mass ratio of Co from 0% to 5%, the certain volume of $Co(NO_3)_2$ aqueous solution is dipped into the prepared g-C_3N_4 aqueous dispersion and kept stirring for 20 h to form homogeneous solution with a water bath at 70 °C for 12 h. After rotoevaporation to dryness, the obtained impregnated sample is annealed at 400 °C for 2 h in a nitrogen atmosphere in the tube furnace and the CoO nanoparticles are in-situ grown on the g-C_3N_4 sheets to obtain CoO/g-C_3N_4 composite.

3.3   Characterization of CoO/g-C$_3$N$_4$ composite

3.3.1   XRD of CoO/g-C$_3$N$_4$ composite

The crystalline structures and the phase components of as-prepared CoO/g-C$_3$N$_4$ composite and g-C$_3$N$_4$ are studied by XRD. As shown in Figure 3.1, the base materials give two typical diffraction peaks at 13.0° and 27.4°, which can be indexed to the (100) and (002) reflections of g-C$_3$N$_4$, respectively. It is assumed that g-C$_3$N$_4$ has a wrinkled sheet-like structure with a relatively smooth surface[24]. For CoO/g-C$_3$N$_4$ composite, the diffraction peaks of g-C$_3$N$_4$ are observed clearly, indicating that these prepared samples maintain the basic structure of g-C$_3$N$_4$. But in comparison with pure g-C$_3$N$_4$, there is a distinct diffraction peak at 36.4°, which can perfectly match the face-centered cubic CoO structure (JCPDS 71-1178). The characteristic diffraction peaks of both CoO and g-C$_3$N$_4$ reveal the successful fabrication of CoO/g-C$_3$N$_4$ composite by in-situ growing of CoO nanoparticles on g-C$_3$N$_4$.

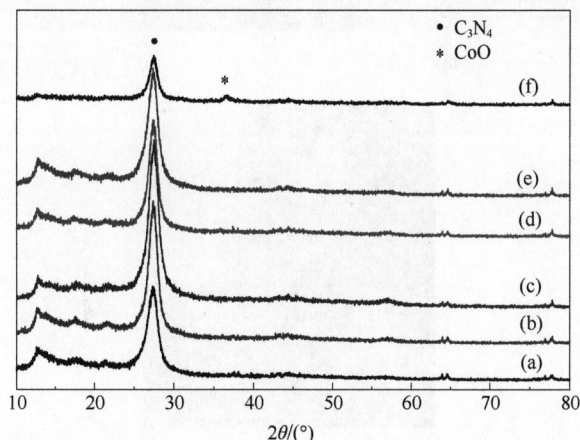

Figure 3.1   XRD patterns of (a) g-C$_3$N$_4$, (b) 0.3% CoO/g-C$_3$N$_4$, (c) 0.6% CoO/g-C$_3$N$_4$, (d) 0.8% CoO/g-C$_3$N$_4$, (e) 1% CoO/g-C$_3$N$_4$ and (f) 5% CoO/g-C$_3$N$_4$ composite

3.3.2 TEM of CoO/g-C₃N₄ composite

The TEM images of the prepared materials are presented in Figure 3.2. As shown in Figure 3.2(b) and Figure 3.2(c), the CoO nanoparticles are highly dispersed by in-situ growing onto the g-C$_3$N$_4$ matrix. From the enlarged high resolution TEM of 5% CoO/g-C$_3$N$_4$ in Figure 3.2(c）, the exposed crystal planes of the obtained CoO can be easily observed, and the lattice fringes with a spacing of 0.25 nm is attributed to the (111) planes of CoO. Based on the XRD and TEM characterizations, it can provide a solid evidence for the successful formation of a CoO/g-C$_3$N$_4$ heterostructure with the in-situ growing method.

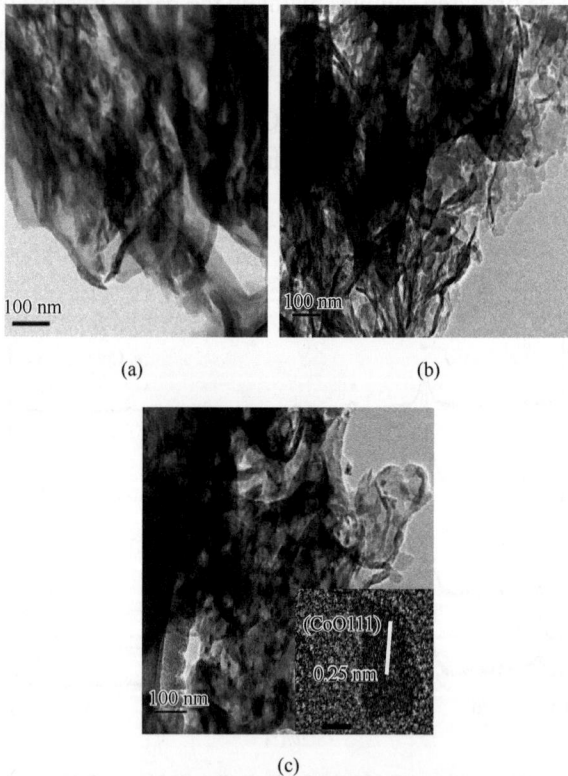

(a)   (b)

(c)

Figure 3.2   TEM and HR-TEM images of (a) g-C$_3$N$_4$, (b) 0.6% CoO/g-C$_3$N$_4$ and (c) 5% CoO/g-C$_3$N$_4$ composite

3.3.3   XPS of CoO/g-C$_3$N$_4$ composite

Surface chemical states of elements and the specific bonding in the prepared samples are in-depth characterized by XPS, and the results are shown in Figure 3.3(a)-(e). The full survey spectrum of the prepared material is shown in Figure 3.3(a). There are three sharp peaks with binding energy values of 285 eV, 399 eV, 530 eV and 782 eV attributed to the signals of C 1s, N 1s, O 1s and Co 2p, respectively in the as-prepared samples. The C 1s spectra[Figure 3.3（b）]can be deconvoluted into three peaks at 284.8 eV, 288.1 eV, and 293.6 eV, respectively. The binding energy at 284.8 eV can be ascribed to the signals of C—C coordination of the surface adventitious carbon. The binding energy at 288.1 eV is attributed to the sp^2-hybridized carbon in N=C—N coordination, while the peak observed at 293.6 eV results from π-excitation[25]. Figure 3.3(c) is presented the N 1s spectra, which can be subdivided into four peaks at 398.7 eV, 399.4 eV, 400.9 eV, and 404.7 eV. The binding energy at 398.7 eV is ascribed to the sp^2-hybridized nitrogen atoms in C=N—C groups[26].The binding energy at 399.4 eV corresponds to the tertiary nitrogen N—C$_3$ groups or H—N—C$_2$[27]. The binding energy at 400.9 eV is results from the amino function groups[26], and the binding energy at 404.7 eV is attributed to charging effects or positive charge localization in heterocycles[28]. The high-resolution XPS spectra of Co 2p for 0.6% CoO/g-C$_3$N$_4$ and 1% CoO/g-C$_3$N$_4$ are displayed in Figure 3.3(e). The weak and diffused Co 2p peaks of 0.6% CoO/g-C$_3$N$_4$ at two pairs of individual peaks centered at 780.3 eV and 796.2 eV confirm the existence of Co, which are identified as the major binding energies of Co^{2+} in CoO[29]. Two peaks at 780.6 eV and 796.5 eV can be attributed to the Co 2p$_{3/2}$ and Co 2p$_{1/2}$ spin-orbit peaks of CoO, respectively. The O 1s spectra with two peaks at about 529 eV and 532.2 eV are shown in Figure 3.3(d). The binding energy at 529 eV is ascribed to the Co-O bond in the CoO phase[30], while the strong peak at about 532.2 eV corresponds to the Co-O-C bond, indicating that a strong interaction exists between CoO and g-C$_3$N$_4$[20]. It can be seen that the signal of the Co-O-C bond becomes more pronounced with the increase of Co loading due to the change of the electronic state of adsorbed oxygen species[31]. Therefore, the interfacial interaction between CoO and g-C$_3$N$_4$ could be greatly enhanced due to the interfacial hybridized Co-O-C bond.

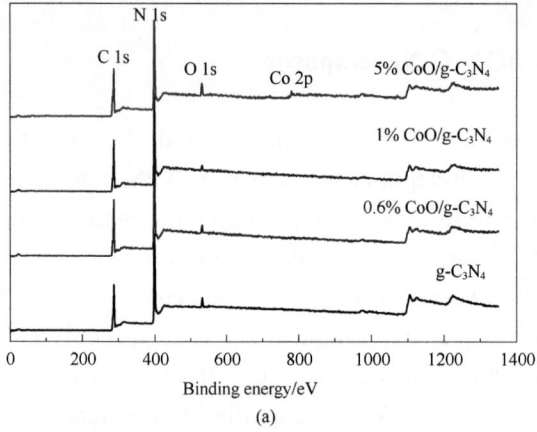

(a)

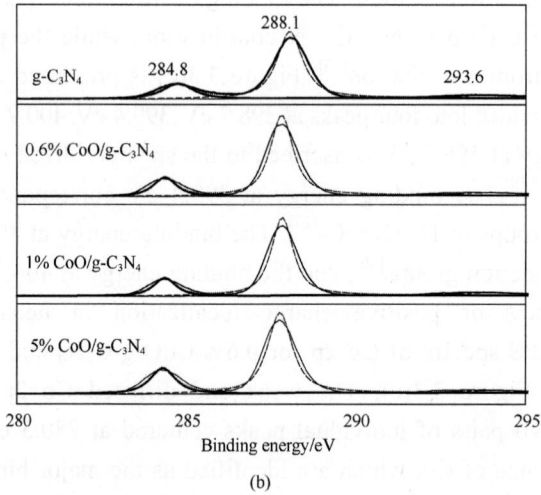

(b)

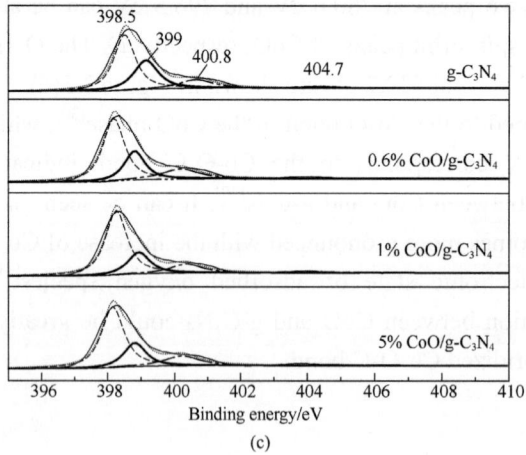

(c)

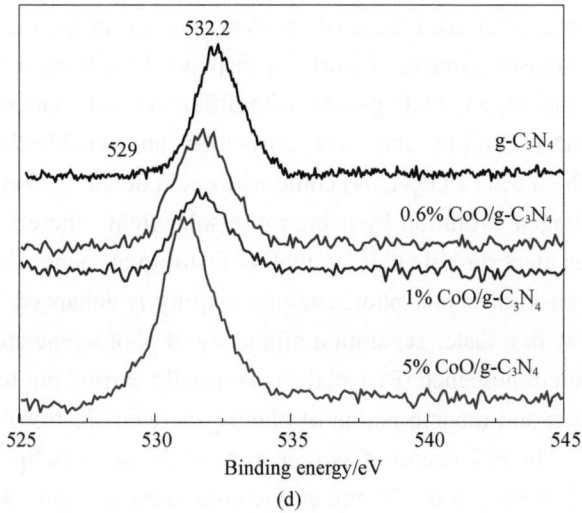

(d)

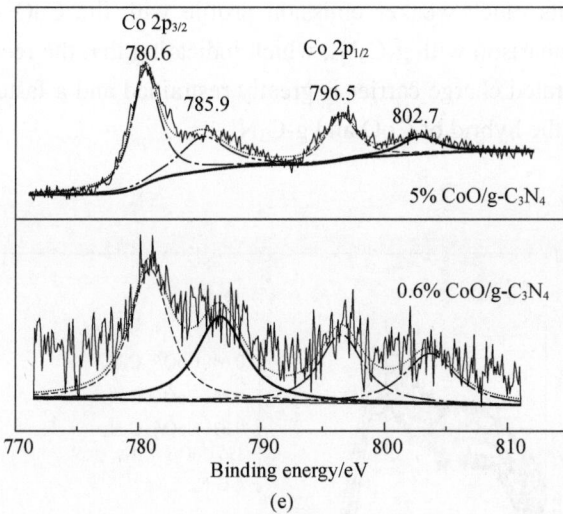

(e)

Figure 3.3   XPS profiles of (a) survey, (b) C 1s, (c) N 1s, (d) O 1s and (e) Co 2p of the prepared samples

3.3.4   UV-vis absorption and PL spectra of CoO/g-C₃N₄ composite

It is widely accepted that the optical and photoelectric properties are of great significance on the photocatalytic activity. UV-Vis absorption spectra and photoluminescence (PL) are measured to identify these properties of g-C_3N_4 and CoO/g-C_3N_4 composite samples. Figure 3.4 displays the UV-vis diffuse reflectance spectra of pure g-C_3N_4 and CoO/g-C_3N_4 with different CoO contents. It is seen that 0.6% CoO/g-C_3N_4 exhibits the best ultraviolet and visible-light absorbance, indicating that the 0.6% CoO/g-C_3N_4 composite could obtain the best photocatalytic activity for hydrogen evolution by using more solar light. The efficient separation of photo-induced electron-hole pairs is another factor to influence the photocatalytic activity. It is well known that photocatalytic activity is enhanced by interfaces of heterojunctions with a faster separation efficiency of photo-generated electron-hole pairs. The photoluminescence (PL) analysis is usually carried out to investigate the transfer, migration and recombination of photo-generated electron-hole pairs of the photocatalyst[20]. The PL spectra of as-prepared g-C_3N_4 and CoO/g-C_3N_4 composite with excitation wavelength of 370 nm are demonstrated in Figure 3.5. CoO/g-C_3N_4 composite exhibits much weaker emission profile with the CoO loading content increasing in comparison with g-C_3N_4, which indicating that the recombination rate of the photogenerated charge carrier is greatly restrained and a faster photoelectron transfer between the hybrid of CoO and g-C_3N_4.

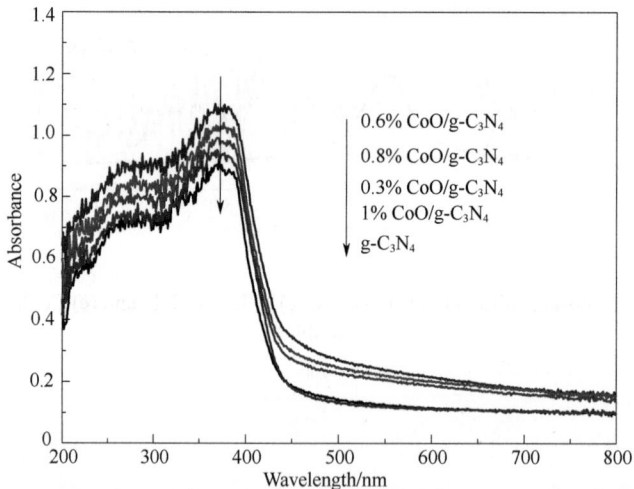

Figure 3.4　UV-vis absorption spectra of g-C_3N_4, 0.3% CoO/g-C_3N_4, 0.6% CoO/g-C_3N_4, 0.8% CoO/g-C_3N_4 and 1% CoO/g-C_3N_4 composite

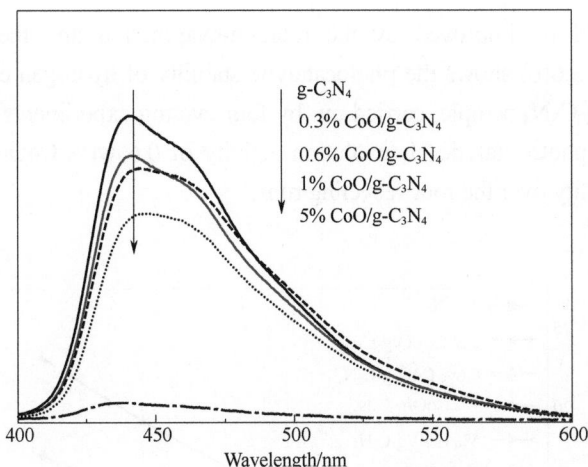

Figure 3.5   PL spectra (λ_{ex} = 370 nm) for the prepared g-C₃N₄, 0.3% CoO/g-C₃N₄, 0.6% CoO/g-C₃N₄, 1% CoO/g-C₃N₄ and 5% CoO/g-C₃N₄ composite

3.4   Photocatalytic H₂ generation testing of CoO/g-C₃N₄ composite

3.4.1   Hydrogen production efficiency

The photocatalytic H₂ evolution performance of the prepared CoO/g-C₃N₄ composite with different CoO content is measured using Pt as a co-catalyst and the results are shown in Figure 3.6. In Figure 3.6(a), it can be found that the photocatalytic H₂ evolution amount for the CoO/g-C₃N₄ composite with 0, 0.3wt%, 0.6wt%, 1wt%, 5wt% and 100 wt% CoO loading content is recorded to be 14.79mmol/g, 17.19mmol/g, 23.25mmol/g, 13.02mmol/g, 1.90mmol/g and 0.019 mmol/g after 5 h, respectively. The photocatalytic H₂ evolution amount increases as the CoO content increases from 0 to 0.6 wt% and then exhibits a decrease with a higher CoO content. This decrease is possibly due to excessive CoO aggregation and the decrease of g-C₃N₄ surface active sites. The 0.6 wt% CoO/g-C₃N₄ composite exhibits the best photocatalytic performance with an average hydrogen evolution rate of 4.65 mmol/(h·g), which is about 57% higher than that of pure g-C₃N₄. Compared with the reported 0.5 wt% CoO/g-C₃N₄ [0.65 mmol/(h·g)][29], 30 wt% CoO/g-C₃N₄ (2.51 μmol/h)[20] and 10 wt% CoO/g-C₃N₄ (0.46 μmol/h)[19], the photocatalytic hydrogen evolution performance of the CoO/g-C₃N₄ composite could be improved by the rotation-evaporation and thermal annealing method. Figure 3.6(b) shows the photocatalytic stability of hydrogen evolution for the 0.6 wt% CoO/g-C$_3$N$_4$ sample, carried out by four cycling experiments under the same condition. The photocatalytic H$_2$ evolution activity of 0.6 wt% CoO/g-C$_3$N$_4$ exhibits favourable stability over the four recycling runs.

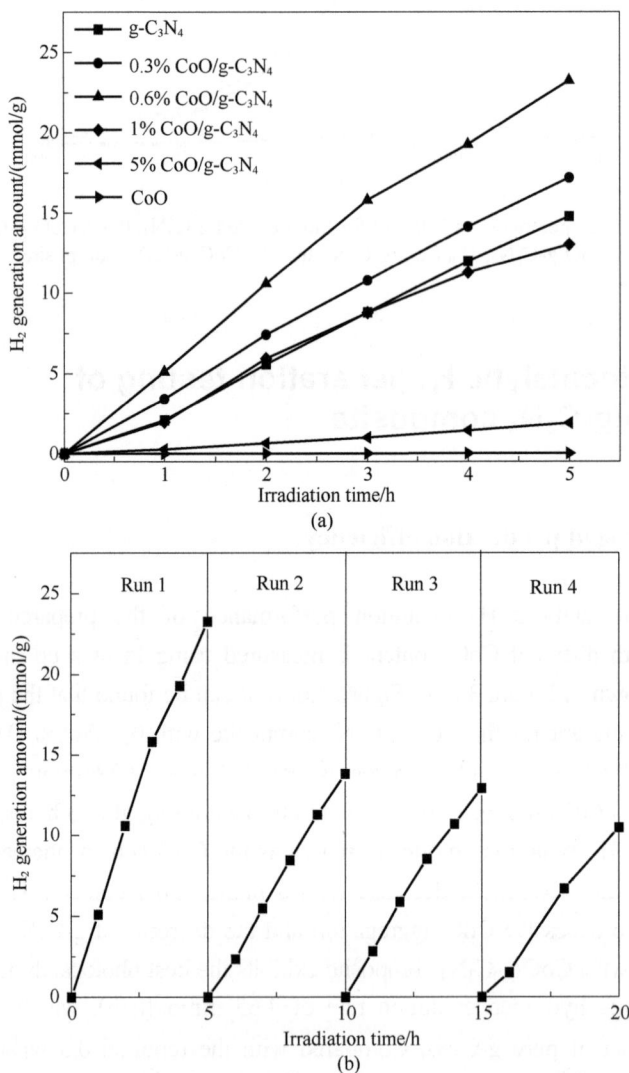

Figure 3.6   (a) The photocatalytic H$_2$ evolution amount of the samples, (b) recyclability of 0.6 wt% CoO/g-C$_3$N$_4$ photocatalyst for the photocatalytic H$_2$ evolution, with 10 vol% TEOA, 1.5 wt% Pt

In the following experiments, as 0.6 wt% CoO/g-C$_3$N$_4$ exhibits the superior photocatalytic H$_2$ evolution activity, its H$_2$ evolution performance for splitting pure water is investigated and the results are shown in Figure 3.7. It is found that 0.6wt% CoO/g-C$_3$N$_4$ can split pure water to generate H$_2$ without sacrificial agent and noble metal Pt, while the pure g-C$_3$N$_4$ and bulk CoO exhibit negligible photocatalytic activity towards H$_2$ generation under the same reaction condition. H$_2$O$_2$ is more easily generated than O$_2$ due to the more positive H$_2$O$_2$/H$_2$O (1.78 V vs. RHE) than O$_2$/H$_2$O (1.23 V vs. RHE)[20]. The drawbacks of the rapid rate of photo-generated electron-hole pairs and being seriously poisoned by H$_2$O$_2$ generated during the reaction of g-C$_3$N$_4$ are the main reasons for inactivation[7]. The photocatalytic H$_2$ evolution amount of 0.6% CoO/g-C$_3$N$_4$ reaches 0.39 mmol/g and the photocatalytic H$_2$ evolution rate is very slow after 2 h under light irradiation. Based on these results, it can be safely concluded that g-C$_3$N$_4$ doped with 0.6 wt% CoO could effectively separate the photo-generated electron-hole pairs to generate H$_2$ from pure water splitting, but it is likely to be greatly poisoned by H$_2$O$_2$ generation during the photocatalytic reaction, causing rapid inactivation.

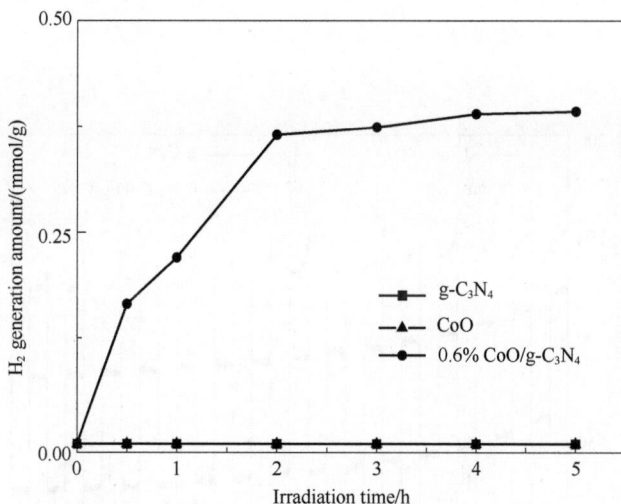

Figure 3.7   H$_2$ evolutions from pure water with g-C$_3$N$_4$, CoO and 0.6 wt% CoO/g-C$_3$N$_4$

3.4.2   Charge separation and transfer performance

The photogenerated charge transfer and separation properties in the 0.6wt%

CoO/g-C$_3$N$_4$ sample are further studied by photoelectrochemical measurements. Electrochemical impedance spectroscopy (EIS) and photocurrents were carried out to obtain the photogenerated charge separation and transfer properties. Figure 3.8(a) shows the transient photocurrent responses for g-C$_3$N$_4$ and 0.6 wt% CoO/g-C$_3$N$_4$ samples with an interval light on/off cycle mode. It can be clearly observed that the transient photocurrent response of 0.6 wt% CoO/g-C$_3$N$_4$ composite is higher than that of pure g-C$_3$N$_4$ sample, suggesting that the interfacial electron transfer and electron-hole separation process between CoO and g-C$_3$N$_4$ is highly efficient. The EIS measurements have been carried out to evaluate the photogenerated electron transfer in CoO/g-C$_3$N$_4$. Figure 3.8(b) presents the experimental Nyquist impedance plots for g-C$_3$N$_4$ and 0.6wt% CoO/g-C$_3$N$_4$ sample in the dark. The Nyquist plot of 0.6wt% CoO/g-C$_3$N$_4$ suggests an apparently smaller arc diameter than that of g-C$_3$N$_4$, indicating that the 0.6wt% CoO/g-C$_3$N$_4$ system can efficiently migrate interfacial charge and separate the photogenerated electronhole pairs at the heterojunction interface across the electrode/electrolyte in agreement with the results of PL result, contributing to the enhancement of photocatalytic hydrogen evolution activity[20].

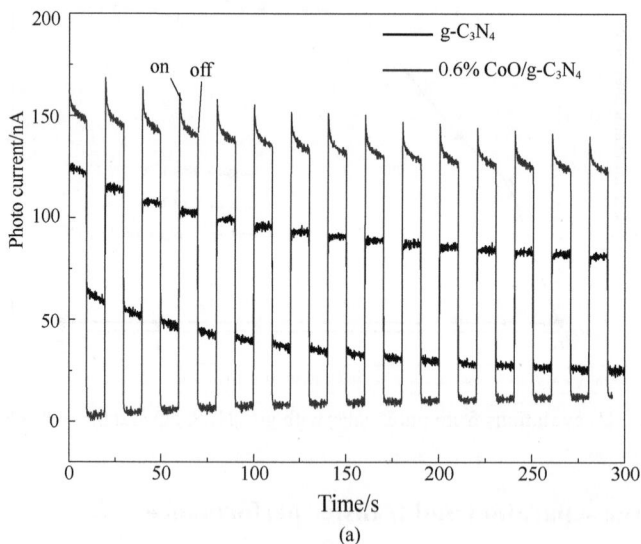

(a)

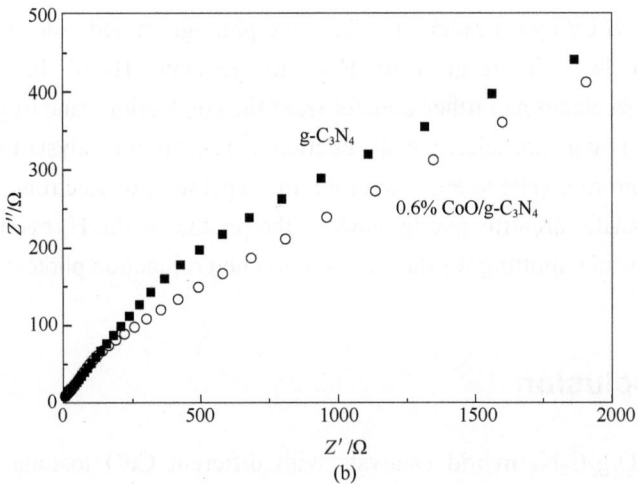

(b)

Figure 3.8   (a) Transient photocurrents, (b) electrochemical impedance spectra of g-C₃N₄ and 0.6
wt% CoO/g-C₃N₄ electrodes at 0.3 V and -0.4 V versus Ag/AgCl

3.4.3   Mechanism of hydrogen production

Based on the above experimental results, a possible mechanism of photocatalytic
H_2 evolution for the CoO/g-C₃N₄ hybrid system is proposed, as shown in Figure 3.9.
First, the electron-hole pairs[32,33] are generated in the conduction band[34] and valence

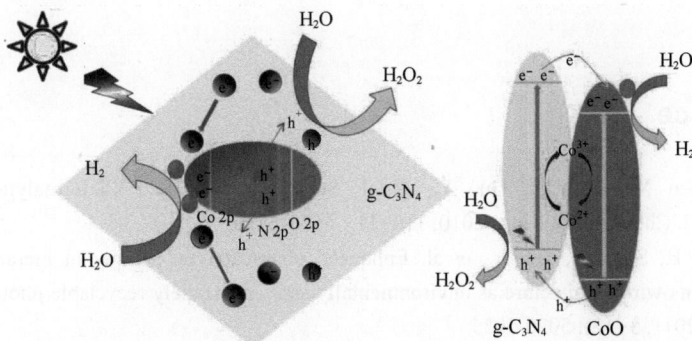

Figure 3.9   Schematic illustrations of the proposed photocatalytic H_2 evolution mechanism for
overall water splitting over the CoO/g-C₃N₄ hybrid catalyst

band of g-C$_3$N$_4$ by light irradiation. Then, the photogenerated holes in the valence band of g-C$_3$N$_4$ will react with H$_2$O to generate H$_2$O$_2$. In contrast, the photogenerated electrons further transfer from the conduction band of g-C$_3$N$_4$ to the surface of CoO nanoparticles, which function as reduction catalysts to catalyze the reduction of protons (H$^+$) to H$_2$. Therefore, the separation of electron-hole pairs and the charge transfer are effective to enhance the photocatalytic H$_2$ evolution activity from overall water splitting for the CoO/g-C$_3$N$_4$ heterojunction photocatalyst.

3.5   Conclusion

The CoO/g-C$_3$N$_4$ hybrid catalysts with different CoO loading contents are facilely prepared to study the photocatalytic H$_2$ evolution activity. The CoO/g-C$_3$N$_4$ composite with 0.6 wt% Co loading shows the highest photocatalytic activity, with an H$_2$ evolution amount of 23.25 mmol/g after 5 h, which is 57% higher than that of g-C$_3$N$_4$. The remarkably enhanced photocatalytic performance for H$_2$ evolution of the CoO/g-C$_3$N$_4$ composite is mainly due to the faster transfer of interfacial charge and more effective separation of electron-hole pairs. The photocatalytic H$_2$ evolution amount of 0.6% CoO/g-C$_3$N$_4$ reaches 0.39 mmol/g by overall water splitting without sacrificial agent and noble metal. But the photocatalytic H$_2$ evolution rate of 0.6 wt% CoO/g-C$_3$N$_4$ is very slow after 2 h because it is easily poisoned by H$_2$O$_2$ generation during the photocatalytic reaction, causing rapid inactivation. In future, CoO/g-C$_3$N$_4$ material with stable photocatalytic H$_2$ evolution activity will be further designed to prevent H$_2$O$_2$ poisoning.

Reference

[1] Chen X., Shen S., Guo L., et al. Semiconductor-based photocatalytic hydrogen generation[J]. Chemical Reviews, 2010, 110 (11): 6503-6570.

[2] Xu F., Shen Y., Sun L., et al. Enhanced photocatalytic activity of hierarchical ZnO nanoplate-nanowire architecture as environmentally safe and facilely recyclable photocatalyst[J]. Nanoscale, 2011, 3 (12): 5020-5025.

[3] Martin D. J., Qiu K., Shevlin S. A., et al. Highly efficient photocatalytic H$_2$ evolution from water using visible light and structure-controlled graphitic carbon nitride[J]. Angewandte Chemie International Edition, 2014, 53 (35): 9240-9245.

[4] Zhang Y., Liu J., Wu G., et al. Porous graphitic carbon nitride synthesized via direct polymerization of urea for efficient sunlight-driven photocatalytic hydrogen production[J]. Nanoscale, 2012, 4 (17): 5300-5303.

[5] Fettkenhauer C., Clavel G., Kailasam K., et al. Facile synthesis of new, highly efficient SnO_2/carbon nitride composite photocatalysts for the hydrogen evolution reaction[J]. Green Chemistry, 2015, 17 (6): 3350-3361.

[6] Yan J., Wu H., Chen H., et al. Fabrication of TiO_2/C_3N_4 heterostructure for enhanced photocatalytic Z-scheme overall water splitting[J]. Applied Catalysis B: Environmental, 2016, 191: 130-137.

[7] Fu Y., Liu C. a., Zhu C., et al. High-performance NiO/g-C_3N_4 composites for visible-light-driven photocatalytic overall water splitting[J]. Inorganic Chemistry Frontiers, 2018, 5 (7): 1646-1652.

[8] Chen J., Shen S., Guo P., et al. In-situ reduction synthesis of nano-sized Cu_2O particles modifying g-C_3N_4 for enhanced photocatalytic hydrogen production[J]. Applied Catalysis B: Environmental, 2014, 152-153, 335-341.

[9] Zou W., Deng B., Hu X., et al. Crystal-plane-dependent metal oxide-support interaction in CeO_2/g-C_3N_4 for photocatalytic hydrogen evolution[J]. Applied Catalysis B: Environmental, 2018, 238: 111-118.

[10] Li J., Yin Y., Liu E., et al. In situ growing Bi_2MoO_6 on g-C_3N_4 nanosheets with enhanced photocatalytic hydrogen evolution and disinfection of bacteria under visible light irradiation[J]. Journal of Hazardous Materials, 2017, 321: 183-192.

[11] Shi J., Cheng C., Hu Y., et al. One-pot preparation of porous Cr_2O_3/g-C_3N_4 composites towards enhanced photocatalytic H_2 evolution under visible-light irradiation[J]. International Journal of Hydrogen Energy, 2017, 42 (7): 4651-4659.

[12] Zhao Z., Sun Y., Dong F., Graphitic carbon nitride based nanocomposites: a review[J]. Nanoscale, 2015, 7 (1): 15-37.

[13] Xu J., Li X., Ju Z., et al. Visible-light-driven overall water splitting boosted by tetrahedrally coordinated blende cobalt(II) oxide atomic layers[J]. Angewandte Chemie, 2019, 131 (10): 3064-3068.

[14] Qin Y., Wang G., Wang Y. Study on the photocatalytic property of La-doped CoO/$SrTiO_3$ for water decomposition to hydrogen[J]. Catalysis Communications, 2007, 8 (6): 926-930.

[15] Zhang J., Yu Z., Gao Z., et al. Porous TiO_2 nanotubes with spatially separated platinum and CoO_x cocatalysts produced by atomic layer deposition for photocatalytic hydrogen production[J]. Angewandte Chemie International Edition, 2017, 56 (3): 816-820.

[16] Shi W., Guo F., Wang H., et al. New insight of water-splitting photocatalyst: H_2O_2-resistance poisoning and photothermal deactivation in sub-micrometer CoO octahedrons[J]. ACS Applied Materials & Interfaces, 2017, 9 (24): 20585-20593.

[17] Shi W., Guo F., Zhu C., et al. Carbon dots anchored on octahedral CoO as a stable visible-light-responsive composite photocatalyst for overall water splitting[J]. Journal of Materials Chemistry A, 2017, 5 (37):19800-19807.

41

[18] Neațu Ș., Puche M., Fornés V., et al. Cobalt-containing layered or zeolitic silicates as photocatalysts for hydrogen generation[J]. Chemical Communications, 2014, 50 (93): 14643-14646.

[19] Han M., Wang H., Zhao S., et al. One-step synthesis of CoO/g-C_3N_4 composites by thermal decomposition for overall water splitting without sacrificial reagents[J]. Inorganic Chemistry Frontiers, 2017, 4 (10): 1691-1696.

[20] Guo F., Shi W., Zhu C., et al. CoO and g-C_3N_4 complement each other for highly efficient overall water splitting under visible light[J]. Applied Catalysis B: Environmental, 2018, 226: 412-420.

[21] Wang N., Li X. Facile synthesis of CoO nanorod/C_3N_4 heterostructure photocatalyst for an enhanced pure water splitting activity[J]. Inorganic Chemistry Communications, 2018, 92: 14-17.

[22] Yatsuya S., Tsukasaki Y., Mihama K., et al. Preparation of extremely fine particles by vacuum evaporation onto a running oil substrate[J]. Journal of Crystal Growth, 1978, 45: 490-494.

[23] Dong F., Wu L., Sun Y., et al. Efficient synthesis of polymeric g-C_3N_4 layered materials as novel efficient visible light driven photocatalysts[J]. Journal of Materials Chemistry, 2011, 21 (39): 15171-15174.

[24] Yang X., Chen Z., Xu J., et al. Tuning the morphology of g-C_3N_4 for improvement of Z-scheme photocatalytic water oxidation[J]. ACS Applied Materials & Interfaces, 2015, 7 (28): 15285-15293.

[25] Li M., Zhang L., Wu M., et al. Mesostructured CeO_2/g-C_3N_4 nanocomposites: remarkably enhanced photocatalytic activity for CO_2 reduction by mutual component activations[J]. Nano Energy, 2016, 19: 145-155.

[26] Gao D., Xu Q., Zhang J., et al. Defect-related ferromagnetism in ultrathin metal-free g-C_3N_4 nanosheets[J]. Nanoscale, 2014, 6 (5): 2577-2581.

[27] Liu C., Jing L., He L., et al. Phosphate-modified graphitic C_3N_4 as efficient photocatalyst for degrading colorless pollutants by promoting O_2 adsorption[J]. Chemical Communications, 2014, 50 (16): 1999-2001.

[28] Wang S., Li C., Wang T., et al. Controllable synthesis of nanotube-type graphitic C_3N_4 and their visible-light photocatalytic and fluorescent properties[J]. Journal of Materials Chemistry A, 2014, 2 (9): 2885-2890.

[29] Mao Z., Chen J., Yang Y., et al. Novel g-C_3N_4/CoO Nanocomposites with Significantly Enhanced Visible-Light Photocatalytic Activity for H_2 Evolution[J]. ACS Applied Materials & Interfaces, 2017, 9 (14): 12427-12435.

[30] Sun Y., Zhou Y., Zhu C., et al. A Pt-Co_3O_4-CD electrocatalyst with enhanced electrocatalytic performance and resistance to CO poisoning achieved by carbon dots and Co_3O_4 for direct methanol fuel cells[J]. Nanoscale, 2017, 9 (17): 5467-5474.

[31] Carrasco J., López-Durán D., Liu Z., et al. In Situ and Theoretical Studies for the Dissociation of Water on an Active Ni/CeO_2 Catalyst: Importance of Strong Metal-Support Interactions for the Cleavage of O—H Bonds[J]. Angewandte Chemie International Edition, 2015,

54 (13): 3917-3921.

[32] Liang Y., Zhou B., Li N., et al. Enhanced dye photocatalysis and recycling abilities of semi-wrapped TiO₂@carbon nanofibers formed via foaming agent driving[J]. Ceramics International, 2018, 44 (2): 1711-1718.

[33] Xu Z., Li X., Wang W., et al. Microstructure and photocatalytic activity of electrospun carbon nanofibers decorated by TiO₂ nanoparticles from hydrothermal reaction/blended spinning[J]. Ceramics International, 2016, 42 (13): 15012-15022.

[34] Xu Z., Wu T., Shi J., et al. Photocatalytic antifouling PVDF ultrafiltration membranes based on synergy of graphene oxide and TiO₂ for water treatment[J]. Journal of Membrane Science, 2016, 520: 281-293.

Chapter 4

ZnO/g-C₃N₄ with N dopant

4.1 Background

Hydrogen has commonly been recognized as an ideal and promising clean energy in the new century. Photocatalytic water splitting for H_2 evolution has been a hot topic since it was reported that TiO_2 semiconductor could obtain H_2 evolution from photocatalytic water splitting in 1972[1]. Subsequently, many efforts have been done to obtain efficient semiconductor materials for water splitting and environmental governance[2-4].

g-C₃N₄ is considered as promising photocatalytic material for environmental remediation and water splitting to produce H_2 under visible light illumination as a consequence of its medium band gap and excellent physicochemical stability. However, the photocatalytic performance of g-C₃N₄ is restricted by the low utilization of visible light, the relatively high photogenerated electron-hole pairs recombination rate and low electrical conductivity[5-7]. g-C₃N₄ nanocomposite photocatalysts with a heterojunction structure have been developed to promote the photocatalytic activity. Mo et al.[8] prepared MnO_2/g-C₃N₄ heterojunction nanocomposite to realize overall water splitting. The Z-scheme mechanism between MnO_2 and g-C₃N₄ is in favor of the improvement of photocatalytic activity, and the H_2 evolution rate reached 28.0 mmol/(g·h). Chang et al.[9] synthesized $NiCo_2O_4$/g-C₃N₄, and it shows excellent water splitting activity with a H_2 evolution rate of 5.48 mmol/(g·h). The g-C₃N₄ composite with a heterojunction structure has superior photocatalytic performance to that of fresh g-C₃N₄, due to high visible-light absorption, fast charge transportation, and efficient separation of photogenerated electron-hole pairs[10,11].

Zinc oxide (ZnO), as one of the common nano-semiconductors, has been widely used for supercapacitors, gas sensors and photocatalysts, owing to its outstanding electrical, mechanical and optical properties. It is reported that nanostructure ZnO is one of the promising photocatalyst candidates because of its low-cost, non-toxic nature and higher solar absorption compared to TiO_2[12]. However, the visible light absorption of ZnO is limited by its wide band gap. It is reported that elemental dopant[13] and heterojunction[14] are favorable to improve the photocatalytic activity. N dopant is an effective approach to improve the visible-light absorption of ZnO photocatalysts[15]. The N-doped ZnO/g-C₃N₄ composite reported by Kumar et al.[16] was synthesized by physical mixing, and the hybrid composite shows excellent visible-light photocatalytic degradation. It is difficult to obtain the high-disperse composite materials with the close-knit heterojunction by the physical mixing method, and the facile rotation-evaporation and thermal annealing method is efficient for preparing high-disperse g-C₃N₄ based composites[17].

In this book, a nitrogen-rich ZnO/g-C₃N₄ composite was synthesized via a combined rotary evaporation and calcination route under a N_2 atmosphere. This composite demonstrates enhanced visible-light-driven photocatalytic performance for both hydrogen evolution and NO photo-oxidation. Its morphology, chemical composition, as well as optical and photoelectrochemical properties were investigated using a series of characterization techniques. The enhancement mechanisms underlying the photocatalytic activity for water splitting and NO photo-oxidation in this nitrogen-rich ZnO/g-C₃N₄ composite are also discussed.

4.2   Preparation of ZnO/g-C₃N₄ with N dopant

Urea, Zinc nitrate hexahydrate and triethanolamine (TEOA) were provided by Aladdin Industrial Corporation (Shanghai, China). These chemicals were used without further processing.

Firstly, the pure g-C₃N₄ sample was obtained by calcination of urea in the covered crucible at 773 K for 3 h with a heating speed of 10 K/min in the muffle furnace. Then, 200 mg g-C₃N₄ powder and 4 mg/mL zinc nitrate aqueous solution were put into the flask with 50 mL of deionized water and dispersed with sonication. The dispersed mixer was then moved into 70 °C water bath and kept stirring for 24 h. After that, the impregnated sample was obtained by rotavaporating method, and the nitrogen-rich ZnO/g-C₃N₄ composite (0.5%~30% mass ratio of Zn to C₃N₄) was obtained by calcination at 673 K in a N_2 atmosphere for 2 h in the tube furnace. The prepared materials were denoted as NZnOCNx, where x refers to the mass ratio of Zn to C₃N₄ (x = 0.5, 1, 5 and 30), which could be abbreviated as NZnOCN0.5, NZnOCN1,

NZnOCN5 and NZnOCN30. The nitrogen-poor 1% ZnO/g-C$_3$N$_4$ composite denoted as ZnOCN1 was prepared by physically mixing ZnO (the product from zinc nitrate by calcination at 673 K) and g-C$_3$N$_4$ without further treatment for reference.

4.3 Characterization of ZnO/g-C$_3$N$_4$ with N dopant

4.3.1 TEM of ZnO/g-C$_3$N$_4$ with N dopant

The morphologies of the prepared samples are observed by TEM, and the obtained results are exhibited in Figure 4.1. The microscope image of fresh g-C$_3$N$_4$ is shown in Figure 4.1(a). It can be seen that the as prepared g-C$_3$N$_4$ has the nearly transparent feature, which indicates g-C$_3$N$_4$ has ultrathin thickness with multi-layer structure. For the NZnOCN1 composite [Figure 4.1(b)], the multi-layer composite structure is retained, but the surface of composite becomes fluffy and rough with loose and soft aggregates. The HR-TEM images of fresh NZnOCN5 are displayed in Figure 4.1(c) and Figure 4.1(d). It can be observed that the 0.32 nm and 0.25 nm lattice fringes spacing are ascribed to (111) crystalline structures of g-C$_3$N$_4$ and (101) crystalline structures of exposed ZnO in the composite[18], respectively, indicating that the ZnO/g-C$_3$N$_4$ heterojunction composite is prepared successfully. To further demonstrate the formation of the composite, the EDS mapping images of NZnOCN1 composite is shown in Figure 4.1(e). It can be seen that carbon, nitrogen, oxygen and zinc are uniformly distributed over the composites, and it confirms that the ZnO nanoparticles are highly dispersed onto the g-C$_3$N$_4$ sheets.

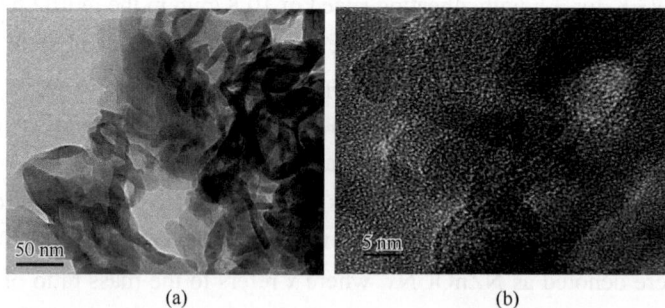

(a)   (b)

(c)   (d)

(e)

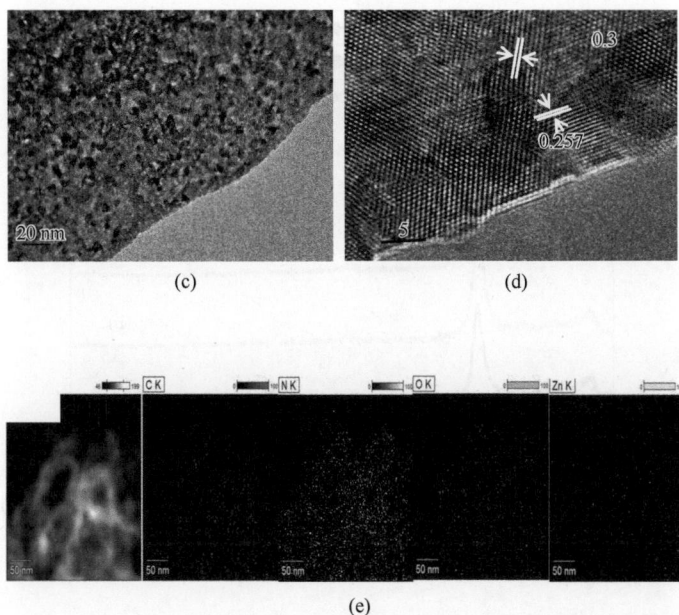

Figure 4.1   TEM and HR-TEM morphology images of (a) g-C₃N₄, (b) NZnOCN1, (c) and (d)
NZnOCN5, (e) The EDS mapping images of NZnOCN1 composite

4.3.2   XRD of ZnO/g-C₃N₄ with N dopant

The XRD patterns of the fresh g-C_3N_4 and nitrogen-rich ZnO/g-C_3N_4 composite are shown in Figure 4.2. From the results of Figure 4.2, it can be seen that the diffraction peaks at 32.4°, 34.0° and 36.3° are attributed to the crystal planes of the ZnO (JCPDS No. 36-1451). With the increase of Zn loading content, the intensity of ZnO diffraction peak becomes more and more apparent. All samples have two obvious diffraction peaks at 27.4° and 13.0° in agreement with the (002) and (100) reflections of carbon nitride[19], indicating that the structure of g-C_3N_4 is not changed by introducing ZnO into the graphitic carbon nitride. The XRD patterns of the nitrogen-rich NZnOCN1 and nitrogen-poor ZnOCN1 composite is displayed in Figure 4.3. As the results shown in Figure S1, the diffraction peak at 27.4° of C_3N_4 (002) in NZnOCN1 composite has a slight blue shift compared with that in ZnOCN1 composite. The blue shift indicates the lattice parameter shrinkage in nitrogen-rich ZnO/g-C_3N_4 composite is caused by N doping.

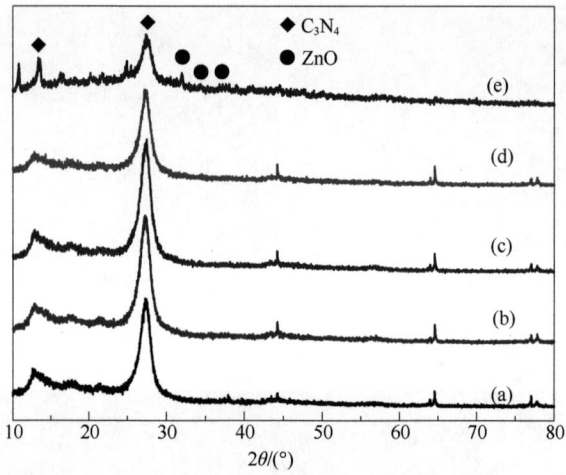

Figure 4.2   XRD patterns of (a) g-C_3N_4, (b) NZnOCN0.5, (c) NZnOCN1, (d) NZnOCN5 and (e) NZnOCN30

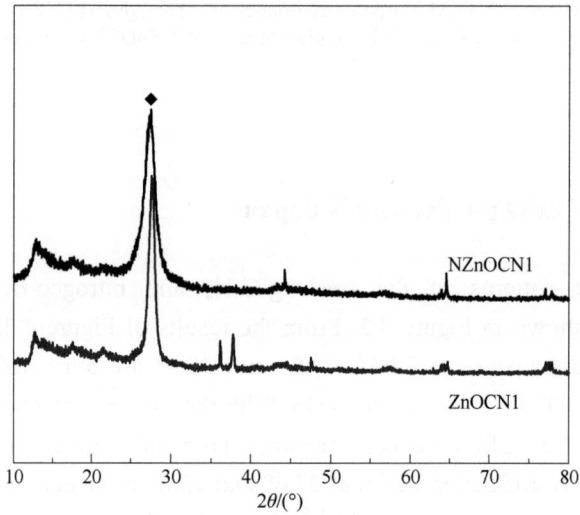

Figure 4.3   XRD patterns of NZnOCN1 and ZnOCN1

4.3.3   XPS of ZnO/g-C_3N_4 with N dopant

In order to determine the nitrogen-rich ZnO/g-C_3N_4 composite, detailed chemical compositions of g-C$_3$N$_4$, NZnOCN1, NZnOCN5 and NZnOCN30 were further investigated by XPS analysis, and the obtained results are exhibited in Figure 4.4. In Figure 4.4(a), three obvious peaks at around 530 eV, 399 eV and 285 eV in the prepared samples could be ascribed to the signals of O 1s, N 1s and C 1s, respectively. It can be seen that the signal of nitrogen in NZnOCN1 is bigger than that of pure g-C$_3$N$_4$, which confirms the nitrogen-rich ZnO/g-C$_3$N$_4$ composite. From the results of atomic relative content in Table 4.1, it is also found that the nitrogen relative content of NZnOCN1 reaches 55.85%, higher than that of carbon nitride, NZnOCN5 and NZnOCN30. N 1s spectra presented in Figure 4.4(b) could be ascribed to four peaks at 398.6 eV, 399.1 eV, 400.7 eV, and 404.6 eV, which are corresponding to the signals of the sp^2-hybridized nitrogen atoms in C=N—C groups[20], N doping[18], the amino function groups[20] and charging effects in heterocycles[21], respectively. The binding energy (BE) and relative content (RC) of N 1s for the prepared samples are shown in Table 4.2. The relative content of N doping of NZnOCN1 composite is about 50% bigger than that of fresh g-C$_3$N$_4$, which determines the rich nitrogen in the prepared ZnO/g-C$_3$N$_4$ composite. It is reported that N doping is more efficiently for altering the band to elevate the photocatalytic H$_2$ evolution activity[22]. As N atoms have stronger electronegativity than that of C atoms, more electrons can transfer to N atoms via N doping bond[22].Therefore, the interfacial heterojunction between ZnO and carbon nitride could be promoted through the rich N doping groups.

The XPS spectra of C 1s of the prepared materials are shown in Figure 4.4(c). There are three peaks of C 1s spectra at around 284.8 eV, 288.1 eV, and 293.6 eV, which can be deconvoluted into the signals of the surface adventitious carbon in C-C coordination, the sp^2-hybridized C in the N=C—N coordination and π-excitation, respectively[17]. The relative content of the sp^2-hybridized carbon in N=C—N coordination of the prepared nitrogen-rich ZnO/g-C$_3$N$_4$ composite is over 70%, which is much higher than that of C-C coordination and π-excitation, as shown in Table 4.3. The XPS spectrum of O 1s is shown in Figure 4.4(d). Two peaks at about 528.9 eV and 532 eV could be corresponding to the lattice oxygen in Zn—O and oxygen atoms in the N—C—O groups, respectively[23].It is found that the binding energy values of O 1s in nitrogen-rich ZnO/g-C$_3$N$_4$ composite are slightly lower than that of g-C$_3$N$_4$ (Table 4.4), which is possible resulted by the heterojunction structure in nitrogen-rich ZnO/g-C$_3$N$_4$ composite. Figure 4.4e displays the XPS spectra of Zn 2p of NZnOCN1, NZnOCN5 and NZnOCN30. Two typical peaks at about 1022 and 1045 eV are respectively ascribed to Zn 2p1 and Zn 2p3, which confirm the existence of the typical Zn^{2+} in ZnO. The relative content of Zn 2p3 in nitrogen-rich $ZnO/g\text{-}C_3N_4$ composite gradually increases with the increase of Zn loading, as shown in Table 4.5.

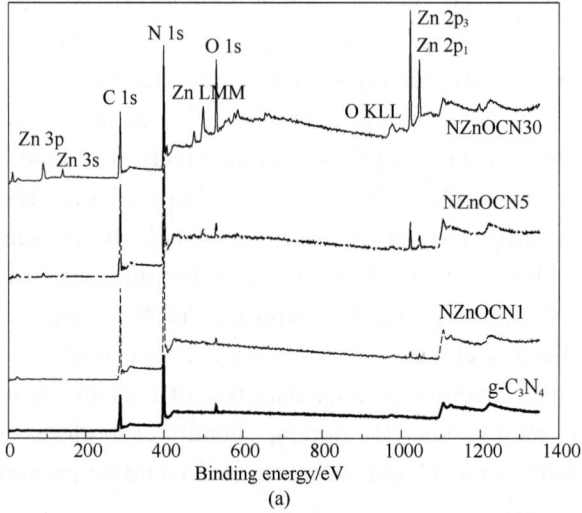

(a)

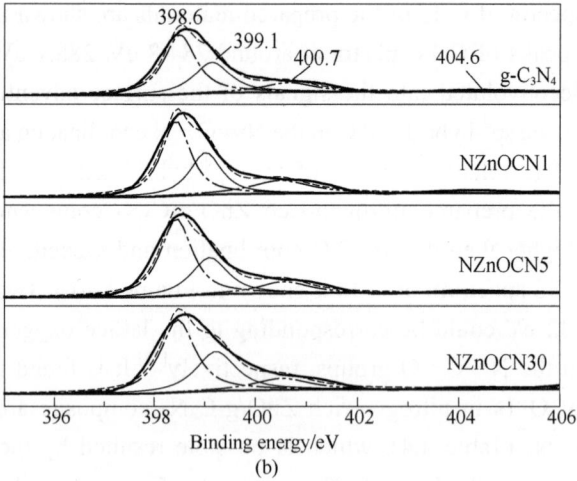

(b)

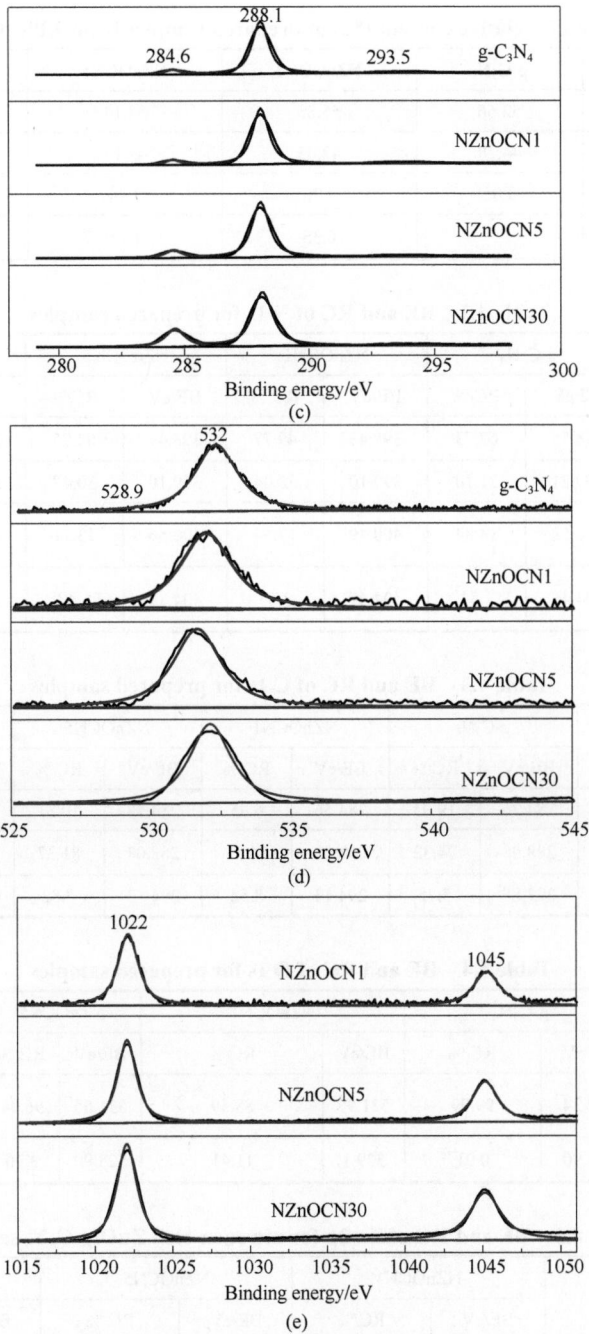

Figure 4.4   (a) XPS survey profiles and high-resolution XPS spectra of (b) N 1s, (c) C 1s, (d) O 1s and (e) Zn 2p of the prepared samples

Table 4.1 Atomic relative content (%) of prepared samples from XPS characterization

Sample	g-C_3N_4	NZnOCN1	NZnOCN5	NZnOCN30
N	52.66	55.85	53.14	43.10
C	44.28	42.03	42.17	37.11
O	3.05	1.76	3.44	15.03
Zn	—	0.35	1.25	4.76

Table 4.2 BE and RC of N 1s for prepared samples

Sample	g-C_3N_4		NZnOCN1		NZnOCN5		NZnOCN30	
	BE/eV	RC/%	BE/eV	RC/%	BE/eV	RC/%	BE/eV	RC/%
N=C—N	398.55	62.73	398.43	49.77	398.44	54.95	398.47	50.41
N doping	399.21	21.10	399.10	30.04	399.10	30.47	399.23	35.65
Amino function	400.58	14.84	400.49	17.54	400.56	13.11	400.52	12.57
Charging effects	404.46	1.33	404.47	2.65	404.39	1.47	404.99	1.36

Table 4.3 BE and RC of C 1s for prepared samples

Sample	g-C_3N_4		NZnOCN1		NZnOCN5		NZnOCN30	
	BE/eV	RC/%	BE/eV	RC/%	BE/eV	RC/%	BE/eV	RC/%
C-C	284.76	19.93	284.56	6.95	284.58	10.81	284.63	22.37
N=C—N	288.08	74.32	288.08	84.53	288.07	81.37	288.13	71.44
π-excitation	293.88	5.75	294.13	8.52	294.13	7.82	294.60	6.19

Table 4.4 BE and RC of O 1s for prepared samples

Sample	g-C_3N_4		NZnOCN1		NZnOCN5		NZnOCN30	
	BE/eV	RC/%	BE/eV	RC/%	BE/eV	RC/%	BE/eV	RC/%
Lattice oxygen	532.34	99.99	531.99	88.59	531.65	96.04	532.11	98.06
N—C—O	528.90	0.01	529.1	11.41	528.90	3.96	528.84	1.94

Table 4.5 BE and RC of Zn 2p for nitrogen-rich ZnO/g-C_3N_4 samples

Sample	NZnOCN1		NZnOCN5		NZnOCN30	
	BE/eV	RC/%	BE/eV	RC/%	BE/eV	RC/%
Zn $2p^1$	1022.15	62.22	1022.07	59.81	1022.04	56.21
Zn $2p^3$	1045.28	37.78	1045.17	40.19	1045.15	43.79

4.3.4   UV-vis absorption and PL spectra of ZnO/g-C₃N₄ with N dopant

It is well known that the optical and photoelectric properties of photocatalysts have much effect on the photocatalytic performance[24-26]. UV-Vis absorption spectra and photoluminescence are carried out to gain these properties of carbon nitride, NZnOCN1, NZnOCN5 and NZnOCN30. The UV-vis absorption spectra of these prepared materials are exhibited in Figure 4.5(a). It can be seen that NZnOCN1 exhibits the best visible light absorbance, which will enhance the photocatalytic hydrogen evolution activity. This is because the ZnO possesses strong absorption in the UV-Vis light region[27], giving rise to the high light absorbance of NZnOCN composite. The PL spectra for g-C₃N₄, NZnOCN0.5, NZnOCN1, NZnOCN5 and NZnOCN30 composite with an excitation wavelength at 370 nm, which are shown in Figure 4.5(b). In comparison with fresh carbon nitride, the signal of PL emission intensity of nitrogen-rich ZnO/g-C₃N₄ composite gradually decreases with the increase of Zn loading, indicating the separation of photo-induced carriers has been effectively promoted. The weaker intensity of PL emission signifies the faster photoelectron transfer[28]. Therefore the recombination rate between the photogenerated electron and hole could be restrained by building the heterojunction structure in the nitrogen-rich ZnO/g-C₃N₄ composite.

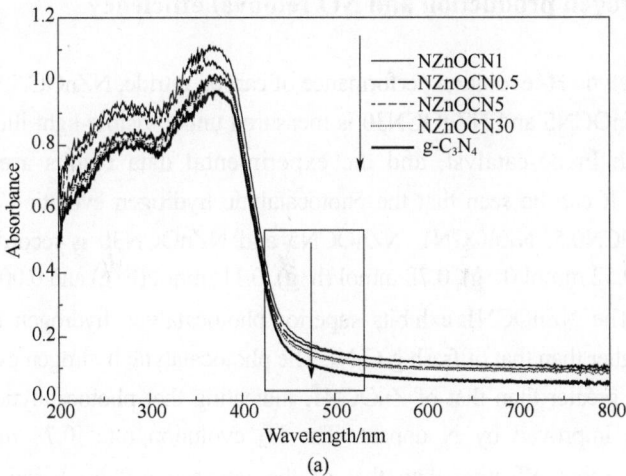

(a)

Figure 4.5

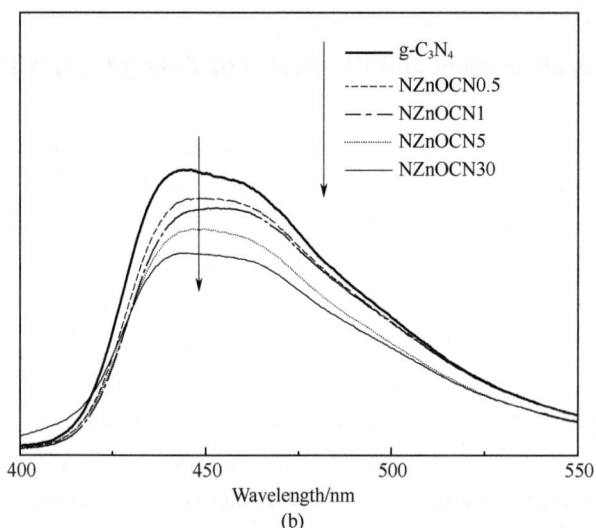

Figure 4.5 (a)UV-vis absorption spectra of the synthesized g-C$_3$N$_4$, NZnOCN0.5, NZnOCN1, NZnOCN5 and NZnOCN30, (b) Photoluminescence (PL) spectra (λ_{ex} = 370 nm) for the synthesized g-C$_3$N$_4$, NZnOCN0.5, NZnOCN1, NZnOCN5 and NZnOCN30 nanocomposite

4.4 Photocatalytic activity testing of ZnO/g-C$_3$N$_4$ with N dopant

4.4.1 Hydrogen production and NO removal efficiency

Photocatalytic H$_2$ evolution performance of carbon nitride, NZnOCN0.5, NZnOCN1, ZnOCN1, NZnOCN5 and NZnOCN30 is measured under visible-light illumination ($\lambda \geqslant$ 400 nm) with Pt co-catalyst, and the experimental data results are exhibited in Figure 4.6(a). It can be seen that the photocatalytic hydrogen evolution rate of carbon nitride, NZnOCN0.5, NZnOCN1, NZnOCN5 and NZnOCN30 is recorded to be 0.44 mmol/(h·g), 0.52 mmol/(h·g), 0.78 mmol/(h·g), 0.11 mmol/(h·g) and 0.004 mmol/(h·g), respectively. The NZnOCN1 exhibits superior photocatalytic hydrogen evolution rate, about 77% higher than that of fresh g-C$_3$N$_4$. The photocatalytic hydrogen evolution rate of NZnOCN1 is greater than that of ZnOCN1, indicating that photocatalytic H$_2$ evolution rate could be improved by N dopant. The H$_2$ evolution rate [0.78 mmol/(h·g)] of NZnOCN1 is more effective than that of the reported g-C$_3$N$_4$ based photocatalytic materials presented in Table 4.6. The photocatalytic H$_2$ evolution stability with four cycling experiments of the NZnOCN1 sample is also measured under visible-light illumination ($\lambda \geqslant 400$ nm) with Pt co-catalyst. In Figure 4.6(b), it is observed that the photocatalytic hydrogen generation activity of NZnOCN1 exhibits well stability after four recycling runs. Therefore it can be safely concluded that the prepared nitrogen-rich ZnO/g-C₃N₄ composite with heterojunction structure indeed exhibits remarkable photocatalytic H₂ evolution activity and stability owing to high visible light absorbance as well as low photogenerated electron-hole pairs recombination rate.

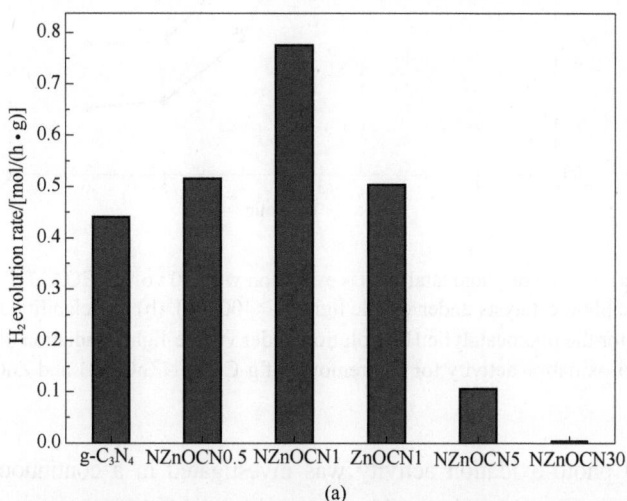

(a)

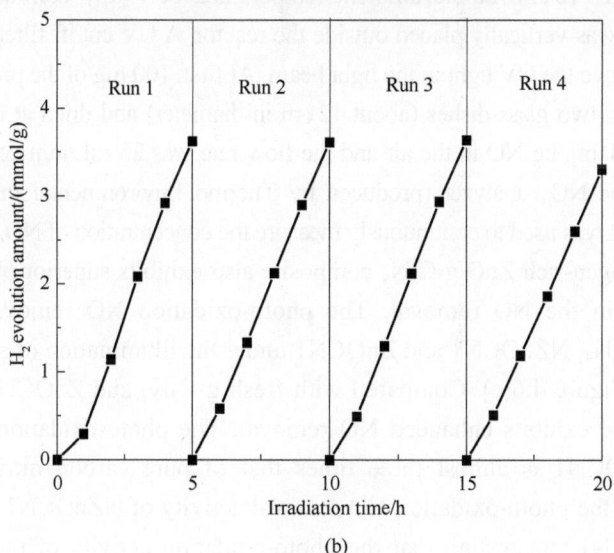

(b)

Figure 4.6

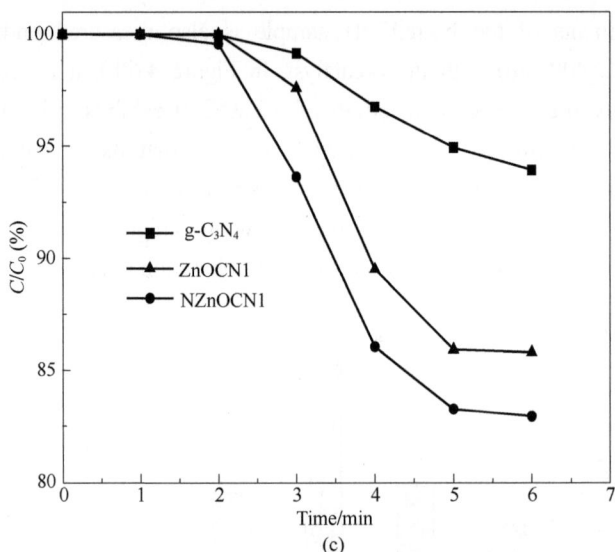

Figure 4.6   (a) the rate of photocatalytic H_2 evolution with 10 vol% TEOA, 1.5 wt% Pt, and 50 mg prepared photocatalysts under visible light ($\lambda \geqslant 400$ nm), (b) recyclability of NZnOCN1 photocatalyst for the photocatalytic H_2 evolution under visible-light irradiation ($\lambda \geqslant 400$ nm), (c) photo-oxidation activity for NO removal of g-C_3N_4, NZnOCN1 and ZnOCN1

The NO photo-oxidation activity was investigated in a continuous-flow reactor (30 cm × 15 cm × 10 cm) at the ambient temperature. A 150 W commercial tungsten halogen lamp was vertically placed outside the reactor. A UV cutoff filter (420 nm) was utilized to remove the UV light in the light beam. At first, 100 mg of the prepared samples were coated on two glass dishes (about 12 cm in diameter) and dried at 60 °C. The gas stream was 600 mg/kg NO in the air and the flow rate was 25 mL/min controlled by the mass flow. The NO_X analyzer (produced by Thermo Environmental Instruments Inc., model 42c-TL) was used to continuously measure the concentration of NO.

The nitrogen-rich ZnO/g-C_3N_4 composite also exhibits superior photo-oxidation performance in the NO removal. The photo-oxidation NO removal rate of the prepared g-C_3N_4, NZnOCN1 and ZnOCN1 under the illumination of simulated light is shown in Figure 4.6(c). Compared with fresh g-C_3N_4 and ZnOCN1, NZnOCN1 nanocomposite exhibits enhanced NO removal. The photo-oxidation NO removal rate of NZnOCN1 is almost three times that of pure carbon nitride. It is also observed that the photo-oxidation NO removal activity of NZnOCN1 is higher than that of ZnOCN1, suggesting that the photo-oxidation activity of the ZnO/g-C_3N_4 composite could   also be enhanced by N dopant.

Table 4.6   Summary of the photocatalytic H₂ evolution rate on g-C₃N₄ based photocatalysts

Photocatalysts	Reaction conditions	Photocatalytic H₂ evolution rate/[mmol/(h·g)]	Reference
NZnOCN1	1.5 wt% Pt, $\lambda \geqslant 400$ nm, 300W	0.78	In this book
NCDs/DCN	RhB solution	0.0037	[29]
g-C₃N₄/ZnO	1 wt% Pt,$\lambda \geqslant 420$ nm, 300W	0.322	[30]
CNT/g-C₃N₄	1.5 wt% Pt, $\lambda \geqslant 400$ nm, 300W	0.17	[31]
Ni₂P/g-C₃N₄	$\lambda \geqslant 420$ nm, 300W	0.47	[32]
NiO/g-C₃N₄	$\lambda \geqslant 420$ nm, 300W	0.03	[33]
CoO/g-C₃N₄	1.5 wt% Pt, 300W	4.65	[19]
GO/g-C₃N₄	1.5 wt% Pt, $\lambda \geqslant 400$ nm, 350W	0.45	[34]
Bi₂MoO₆/g-C₃N₄	3 wt% Pt, $\lambda \geqslant 420$ nm, 300W	0.56	[35]
Carbon/g-C₃N₄	1 wt% Pt, $\lambda \geqslant 420$ nm, 300W	0.21	[6]
Cr₂O₃/g-C₃N₄	5 wt% Pt, $\lambda \geqslant 420$ nm, 300W	0.21	[36]
g-C₃N₄(580)-T	3 wt% Pt, $\lambda \geqslant 420$ nm, 300W	1.39	[26]
PtNi/g-C₃N₄	$\lambda \geqslant 420$ nm, 300W	2.0	[37]
g-C₃N₄/Nb₂O₅	1.5 wt% Pt, $\lambda \geqslant 420$ nm, 300W	1.71	[38]
Fe₂O₃/g-C₃N₄	5 vol% Pt, $\lambda \geqslant 400$ nm, 300W	0.78	[39]
SrTiO₃/g-C₃N₄	1 wt% Pt, $\lambda \geqslant 420$ nm, 300W	0.44	[40]
CDs/g-C₃N₄	3 wt% Pt, $\lambda \geqslant 420$ nm, 300W	2.34	[28]
CuS/g-C₃N₄	$\lambda \geqslant 420$ nm, 300W	0.017	[41]
FeCoP/g-C₃N₄	$\lambda \geqslant 420$ nm, 300W	0.35	[25]
CeO₂/g-C₃N₄	3 wt% Pt, $\lambda \geqslant 420$ nm, 300W	0.86	[42]

4.4.2   Charge separation and transfer performance

In order to determine the enhanced photocatalytic mechanism of nitrogen-rich ZnO/g-C₃N₄ photocatalyst, transient photocurrent responses of prepared materials have been carried out and compared with an interval light on/off cycle mode under visible-light illumination. In Figure 4.7(a), the prepared NZnOCN1 composite exhibits a higher photocurrent and better photo-stability test than that of g-C₃N₄, suggesting improvement of the charge transfer as well as separation process between ZnO and carbon nitride in the nitrogen-rich ZnO/g-C₃N₄ composite. The electrochemical impedance spectroscopy (EIS) is measured to identify the electron transfer properties. The experimental Nyquist impedance plots for carbon nitride, NZnOCN1, and ZnOCN1 composite is exhibited in Figure 4.7(b). Normally, the

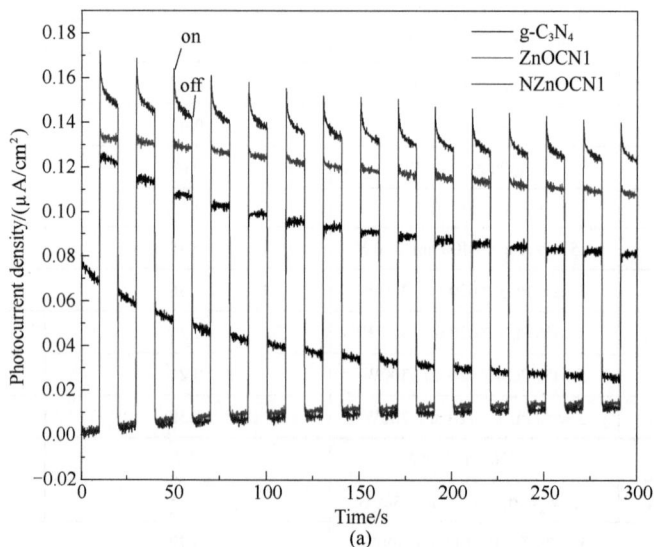

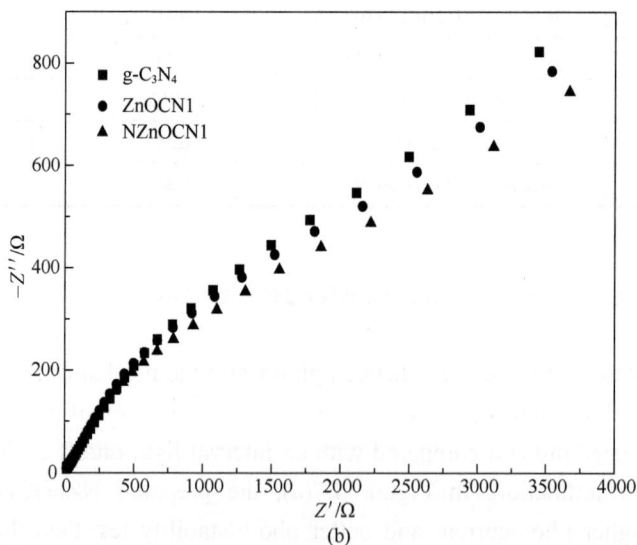

Figure 4.7   (a) Transient photocurrent density of g-C$_3$N$_4$ and nitrogen-rich ZnO/g-C$_3$N$_4$ electrodes at 0.3 V versus Ag/AgCl, (b) electrochemical impedance spectra of the as-prepared samples at -0.4 V versus Ag/AgCl

smaller arc radius on the EIS Nyquist plot, the higher effective separation of photogenerated electron-hole pairs and better interfacial charge transfer ability across the electrode/electrolyte would be[43]. It is evidently seen that the arc radius of the pure g-C₃N₄ electrode is smaller than that of NZnOCN1 electrode, suggesting the NZnOCN1 composite exhibits stronger electronic conductivity to effectively separate the photogenerated electron-hole pairs. Based on the photocurrent and EIS analysis, it is assumed that the photocatalytic hydrogen evolution activity of nitrogen-rich ZnO/g-C₃N₄ photocatalyst could be enhanced by the fast charge transfer and effective separation of photogenerated electron-hole pairs.

4.4.3   Mechanism of enhanced photocatalytic activity

Based on the above experimental results and analysis, the possible mechanism for the enhancement of photocatalytic activity by nitrogen-rich ZnO/g-C₃N₄ composite is proposed in Figure 4.8. Under visible light illumination, the electron-hole pairs will be excited in both ZnO and carbon nitride. It is universally acknowledged that the photocatalytic performance can be enhanced by the efficient separation of photoinduced electrons and holes. The generated electron-hole pairs could migrate via the heterojunction interface of two semiconductor materials with suitable position of VB and CB. The band gap energies of CB and VB of ZnO are around −0.2 eV and +2.84 eV, while the CB and VB potentials of g-C₃N₄ are around −1.22 eV and +1.56 eV, respectively. The CB of g-C₃N₄ is higher than that of ZnO. The photogenerated electrons could be easily transferred from g-C₃N₄ to the CB of ZnO by the potential difference. The electron transfer process will be hastened and the interface energy will be reduced by the heterojunction structure of nitrogen-rich ZnO/g-C₃N₄ composite. For the photocatalytic H₂ generation reaction, the possible mechanism is the Type II mechanism. In this mechanism, the electrons on the CB of g-C₃N₄ will transfer to the CB of ZnO and the holes on the VB of ZnO via N dopant will transfer to the VB of g-C₃N₄. The accumulated electrons in the CB of ZnO could further transfer to Pt nanoparticles to catalyze the reduction of protons to H₂, while the holes in VB of carbon nitride are depleted by the scavenging of TEOA. As the CB potential of ZnO with N doping (−0.2 eV vs. NHE) is more positive than the standard redox potential $E(\cdot O_2/\cdot O_2^-)$ (−0.3 eV vs. NHE), the absorbed $\cdot O_2$ molecules cannot be reduced into $\cdot O_2^-$ by the photogenerated electrons on the CB of ZnO[16]. In Type II mechanism, the electrons and holes will accumulate in the CB of ZnO with N doping and VB of g-C₃N₄, which is unfavorable for the generation of $\cdot O_2^-$. Therefore, the Z-scheme system will be suitable for NO photo-oxidation. In Z-scheme mechanism, the generated electrons on CB of ZnO with N doping could migrate to the VB of g-C_3N_4, thus recombination of electrons and holes could be effectively limited. Besides, the nitrogen-rich ZnO/g-C_3N_4 composite with a heterojunction structure could effectively reduce the migration resistance of photogenerated carriers. The photocatalytic efficiency could be enhanced by the increasing production of $\cdot O_2^-$ due to the accumulation of electrons in the CB of g-C_3N_4.

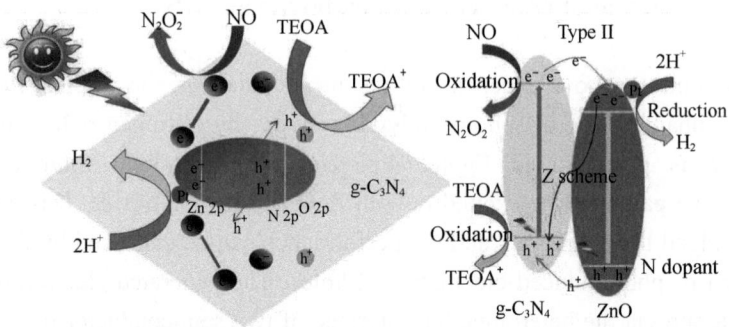

Figure 4.8   Schematic illustrations of the proposed photocatalytic mechanism over the nitrogen-rich ZnO/g-C_3N_4 composite under visible light irradiation

4.5 Conclusion

The nitrogen-rich ZnO/g-C_3N_4 composite with different mass contents of Zn loading was prepared to evaluate visible-light-driven photocatalytic performance for hydrogen generation and NO photo-oxidation. Characterization results confirmed the formation of the nitrogen-rich ZnO/g-C_3N_4 composite. It is verified from the experimental data that the photocatalytic performance for hydrogen generation and NO photo-oxidation of the ZnO/g-C_3N_4 composite could be elevated by N dopant. NZnOCN1 composite exhibits the best photocatalytic activity for H_2 evolution rate of 0.78 mmol/(h·g) under visible light illumination, about 77% higher than that of pure carbon nitride. The photo-oxidation NO removal rate of NZnOCN1 is almost two times higher than that of carbon nitride. As the photogenerated electrons could be easily transferred from g-C_3N_4 to the CB of ZnO by the potential difference, the type II mechanism is the possible mechanism of the photocatalytic H_2 generation reaction. The results show that the Z-scheme mechanisms of nitrogen-rich ZnO/g-C₃N₄ composite are conducive to NO photo-oxidation because the Type II mechanism is unfavorable for the generation of $\cdot O_2^-$.

Reference

[1] Fujishima A., Honda K. Electrochemical photolysis of water at a semiconductor electrode[J]. Nature, 1972, 238: 37.

[2] Xie Z., Liu K., Ren X., et al. Amino-acid-mediated biomimetic formation of light-harvesting antenna capable of hydrogen evolution[J]. ACS Applied Bio Materials, 2018, 1 (3): 748-755.

[3] Peng S., Dan M., Guo F., et al.Template synthesis of ZnIn₂S₄ for enhanced photocatalytic H_2 evolution using triethanolamine as electron donor[J]. Colloids and Surfaces A: Physicochemical and Engineering Aspects, 2016, 504: 18-25.

[4] Liu K., Yuan C., Zou Q., et al.Self-assembled zinc/cystine-based chloroplast mimics capable of photoenzymatic reactions for sustainable fuel synthesis[J]. Angewandte Chemie International Edition, 2017, 56 (27): 7876-7880.

[5] Zhao S., Zhang Y., Zhou Y., et al. Facile one-step synthesis of hollow mesoporous g-C₃N₄ spheres with ultrathin nanosheets for photoredox water splitting[J]. Carbon, 2018, 126: 247-256.

[6] Xu Q., Cheng B., Yu J., et al. Making co-condensed amorphous carbon/g-C₃N₄ composites with improved visible-light photocatalytic H_2-production performance using Pt as cocatalyst[J]. Carbon, 2017, 118: 241-249.

[7] Huang J., Liu J., Yan J., et al.Enhanced photocatalytic H_2 evolution by deposition of metal nanoparticles into mesoporous structure of g-C₃N₄[J]. Colloids and Surfaces A: Physicochemical and Engineering Aspects, 2020, 585: 124067.

[8] Mo Z., Xu H., Chen Z., et al. Construction of MnO₂/Monolayer g-C₃N₄ with Mn vacancies for Z-scheme overall water splitting[J]. Applied Catalysis B: Environmental, 2019, 241: 452-460.

[9] Chang W., Xue W., Liu E., et al. Highly efficient H_2 production over NiCo₂O₄ decorated g-C₃N₄ by photocatalytic water reduction[J]. Chemical Engineering Journal, 2019, 362: 392-401.

[10] Chen X., Shen S., Guo L., et al. Semiconductor-based Photocatalytic Hydrogen Generation[J]. Chemical Reviews, 2010, 110 (11): 6503-6570.

[11] Xu F., Shen Y., Sun L., et al. Enhanced photocatalytic activity of hierarchical ZnO nanoplate-nanowire architecture as environmentally safe and facilely recyclable photocatalyst[J]. Nanoscale, 2011, 3 (12): 5020-5025.

[12] Ong C. B., Ng L. Y., Mohammad A. W. A review of ZnO nanoparticles as solar photocatalysts: Synthesis, mechanisms and applications[J]. Renewable and Sustainable Energy Reviews, 2018, 81: 536-551.

[13] Yu W., Zhang J., Peng T. New insight into the enhanced photocatalytic activity of N-, C- and S-doped ZnO photocatalysts[J]. Applied Catalysis B: Environmental, 2016, 181: 220-227.

[14] Liu J., Yan X.-T., Qin X.-S., et al. Light-assisted preparation of heterostructured g-C$_3$N$_4$/ZnO nanorods arrays for enhanced photocatalytic hydrogen performance[J]. Catalysis Today, 2019.

[15] Oliveira J. A., Nogueira A. E., Gonçalves M. C. P., et al. Photoactivity of N-doped ZnO nanoparticles in oxidative and reductive reactions[J]. Applied Surface Science, 2018, 433: 879-886.

[16] Kumar S., Baruah A., Tonda S., et al. Cost-effective and eco-friendly synthesis of novel and stable N-doped ZnO/g-C$_3$N$_4$ core-shell nanoplates with excellent visible-light responsive photocatalysis[J]. Nanoscale, 2014, 6 (9): 4830-4842.

[17] Liu X., He L., Chen X., et al. Facile synthesis of CeO$_2$/g-C$_3$N$_4$ nanocomposites with significantly improved visible-light photocatalytic activity for hydrogen evolution[J]. International Journal of Hydrogen Energy, 2019, 44 (31): 16154-16163.

[18] Liu Y., Liu H., Zhou H., et al. A Z-scheme mechanism of N-ZnO/g-C$_3$N$_4$ for enhanced H$_2$ evolution and photocatalytic degradation[J]. Applied Surface Science, 2019, 466: 133-140.

[19] Liu X., Zhang Q., Liang L., et al. In-situ growing of CoO nanoparticles on g-C$_3$N$_4$ composites with highly improved photocatalytic activity for hydrogen evolution[J]. Royal Society Open Science, 2019, 6 (7): 190433.

[20] Gao D., Xu Q., Zhang J., et al. Defect-related ferromagnetism in ultrathin metal-free g-C$_3$N$_4$ nanosheets[J]. Nanoscale, 2014, 6 (5): 2577-2581.

[21] Wang S., Li C., Wang T., et al. Controllable synthesis of nanotube-type graphitic C$_3$N$_4$ and their visible-light photocatalytic and fluorescent properties[J]. Journal of Materials Chemistry A ,2014, 2 (9): 2885-2890.

[22] Wang F., Chen P., Feng Y., et al. Facile synthesis of N-doped carbon dots/g-C$_3$N$_4$ photocatalyst with enhanced visible-light photocatalytic activity for the degradation of indomethacin[J]. Applied Catalysis B: Environmental, 2017, 207: 103-113.

[23] Zou W., Shao Y., Pu Y., et al. Enhanced visible light photocatalytic hydrogen evolution via cubic CeO$_2$ hybridized g-C$_3$N$_4$ composite[J]. Applied Catalysis B: Environmental, 2017, 218: 51-59.

[24] An Z., Gao J., Wang L., et al. Novel microreactors of polyacrylamide (PAM)CdS microgels for admirable photocatalytic H$_2$ production under visible light[J]. International Journal of Hydrogen Energy, 2019, 44 (3): 1514-1524.

[25] Cheng L., Xie S., Zou Y., et al. Noble-metal-free Fe$_2$P-Co$_2$P co-catalyst boosting visible-light-driven photocatalytic hydrogen production over graphitic carbon nitride: The synergistic effects between the metal phosphides[J]. International Journal of Hydrogen Energy, 2019, 44 (8): 4133-4142.

[26] Hong Y., Liu E., Shi J., et al. A direct one-step synthesis of ultrathin g-C$_3$N$_4$ nanosheets

from thiourea for boosting solar photocatalytic H_2 evolution[J]. International Journal of Hydrogen Energy, 2019, 44 (14): 7194-7204.

[27] Wang J., Xia Y., Zhao H., et al. Oxygen defects-mediated Z-scheme charge separation in g-C_3N_4/ZnO photocatalysts for enhanced visible-light degradation of 4-chlorophenol and hydrogen evolution[J]. Applied Catalysis B: Environmental, 2017, 206: 406-416.

[28] Wang K., Wang X., Pan H., et al. In situ fabrication of CDs/g-C_3N_4 hybrids with enhanced interface connection via calcination of the precursors for photocatalytic H_2 evolution[J]. International Journal of Hydrogen Energy, 2018, 43 (1): 91-99.

[29] Liu H., Liang J., Fu S., et al. N doped carbon quantum dots modified defect-rich g-C_3N_4 for enhanced photocatalytic combined pollutions degradation and hydrogen evolution[J]. Colloids and Surfaces A: Physicochemical and Engineering Aspects, 2020, 591: 124552.

[30] Zeng D., Ong W.-J., Chen Y., et al. Co_2P nanorods as an efficient cocatalyst decorated porous g-C_3N_4 nanosheets for photocatalytic hydrogen production under visible light irradiation[J]. Particle & Particle Systems Characterization, 2018, 35 (1): 1700251.

[31] Christoforidis K. C., Syrgiannis Z., La Parola V., et al. Metal-free dual-phase full organic carbon nanotubes/g-C_3N_4 heteroarchitectures for photocatalytic hydrogen production[J]. Nano Energy, 2018, 50: 468-478.

[32] Zeng, D., Xu, W., Ong, W. J., Xu, J., Ren, H., et al.,Toward noble-metal-free visible-light-driven photocatalytic hydrogen evolution: monodisperse sub-15nm Ni_2P nanoparticles anchored on porous g-C_3N_4 nanosheets to engineer 0D-2D heterojunction interfaces. Applied Catalysis B: Environmental 2018, 221: 47-55.

[33] Fu Y., Liu C. A., Zhu C., et al. High-performance NiO/g-C_3N_4 composites for visible-light-driven photocatalytic overall water splitting[J]. Inorganic Chemistry Frontiers, 2018, 5 (7): 1646-1652.

[34] Xiang Q., Yu J., Jaroniec M.Preparation and enhanced visible-light photocatalytic H_2-production activity of graphene/C_3N_4 composites[J]. The Journal of Physical Chemistry C, 2011, 115 (15): 7355-7363.

[35] Li J., Yin Y., Liu E., et al. In situ growing Bi_2MoO_6 on g-C_3N_4 nanosheets with enhanced photocatalytic hydrogen evolution and disinfection of bacteria under visible light irradiation[J]. Journal of Hazardous Materials, 2017, 321: 183-192.

[36] Shi J., Cheng C., Hu Y., et al. One-pot preparation of porous Cr_2O_3/g-C_3N_4 composites towards enhanced photocatalytic H_2 evolution under visible-light irradiation[J]. International Journal of Hydrogen Energy, 2017, 42 (7): 4651-4659.

[37] Peng W., Zhang S. S., Shao Y. B., et al. Bimetallic PtNi/g-C_3N_4 nanotubes with enhanced photocatalytic activity for H_2 evolution under visible light irradiation[J]. International Journal of Hydrogen Energy,2018, 43 (49): 22215-22225.

[38] Huang Q. Z., Wang J. C., Wang P. P., et al. In-situ growth of mesoporous Nb_2O_5 microspheres on g-C_3N_4 nanosheets for enhanced photocatalytic H_2 evolution under visible light irradiation[J]. International Journal of Hydrogen Energy, 2017, 42 (10): 6683-6694.

[39] Li Y. P., Li F. T., Wang X. J., et al. Z-scheme electronic transfer of quantum-sized α-Fe_2O_3 modified g-C_3N_4 hybrids for enhanced photocatalytic hydrogen production[J].

International Journal of Hydrogen Energy, 2017, 42 (47): 28327-28336.

[40] Xu X., Liu G., Randorn C., et al. g-C_3N_4 coated $SrTiO_3$ as an efficient photocatalyst for H_2 production in aqueous solution under visible light irradiation[J]. International Journal of Hydrogen Energy ,2011, 36 (21): 13501-13507.

[41] Chen T., Song C., Fan M., et al. In-situ fabrication of CuS/g-C_3N_4 nanocomposites with enhanced photocatalytic H_2-production activity via photoinduced interfacial charge transfer[J]. International Journal of Hydrogen Energy, 2017, 42 (17): 12210-12219.

[42] Zou W., Deng B., Hu X., et al. Crystal-plane-dependent metal oxide-support interaction in CeO_2/g-C_3N_4 for photocatalytic hydrogen evolution[J]. Applied Catalysis B: Environmental, 2018, 238: 111-118.

[43] Guo F., Shi W., Zhu C., et al. CoO and g-C_3N_4 complement each other for highly efficient overall water splitting under visible light[J]. Applied Catalysis B: Environmental, 2018, 226: 412-420.

Chapter 5

Manganese dioxides with different exposed crystal plane supported on g-C$_3$N$_4$

5.1 Background

Manganese oxide, as one of the most abundant rare metal oxides is partially filled with d-levels in favor of the visible light absorption. Manganese oxide has shown its advantages in photocatalytic water splitting owing to its rich oxygen vacancy and well redox ability with couples of Mn^{3+}/Mn^{4+}. It has been reported that MnO$_x$ nanoparticles could be flexibly supported on g-C$_3$N$_4$ could to form Type II structure in many photocatalytic fields. The MnO$_x$/g-C$_3$N$_4$ nano-composite was prepared for NO removal by Dong et al., and the obtained nanocomposite showed better photocatalytic performance compared to the fresh g-C$_3$N$_4$[1]. Wang et al. designed a 2D-2D MnO$_2$/g-C$_3$N$_4$ heterojunction photocatalyst for photocatalytic CO$_2$ reduction, and it is found that the heterojunction photocatalyst showed greatly enhanced photocatalytic activity in the reduction of CO$_2$ compared than fresh g-C$_3$N$_4$ and MnO$_2$[2]. Although photocatalytic activity of MnO$_2$/g-C$_3$N$_4$ has been investigated, some fundamental problems still need to be overcome. As is known to all, various MnO$_2$ nano-materials with different crystal planes show different electronic structure and redox ability. It is reported that the oxygen defect formation energy of a transition metal oxide depends on the crystal plane[3]. Li et al. assumed that the preferred on transition metal oxide with exposed [001] crystal plane is in favor of photo-electrons generation, while an exposed [111] crystal plane would encourage holes to migrate[4]. Xia et al. prepared MnO$_2$ nanocomposite with an exposed [100] crystal plane supported on g-C$_3$N$_4$, exhibiting good photocatalytic activity[5]. Based on that, MnO_2 with different exposed crystal planes supported on the interface of g-C_3N_4 would have electron-hole separation abilities, resulting in the different photocatalytic water-splitting activities.

In this book, MnO_2 with different planes exposed were prepared on g-C_3N_4 to investigate the photocatalytic hydrogen production activity under visible light. Furthermore, a series of characterizations were employed to reveal the influence on photocatalytic activity by MnO_2 with different exposed planes. Finally, the enhanced mechanism of photocatalytic hydrogen production activity from water splitting was proposed.

5.2　Preparation of MnO_2/g-C_3N_4 composite

Urea, manganese acetate tetrahydrate, potassium permanganate, manganese nitrate and triethanolamine (TEOA) were purchased from Aladdin Industrial Corporation (Shanghai, China). These chemicals were used without further processing.

(1) g-C_3N_4 preparation

Firstly, the pure g-C_3N_4 material was synthesized by heating at 500 °C for 3 h in the covered crucible. Then, the prepared g-C_3N_4 was put in 50 mL pure water and kept stirring for 20 h at 70 °C. The products were obtained by rotavaporating and drying at 120 °C for further use.

(2) Preparation of MnO_2/g-C_3N_4 composite by precipitation

A certain amount of manganese acetate tetrahydrate was added into 100 mL of deionized water and dissolved by stirring at room temperature. The prepared 1 mol/L manganese nitrate solution and a certain amount of g-C_3N_4 was added into a certain amount of potassium permanganate solution and the solution was stirred for 24 hours. Then, the precipitate was filtered and dried at 120 °C for 12 hours. After grinding, the MnO_2/g-C_3N_4 composite prepared by precipitation, named p-MnOCN, was obtained by heating at 400 °C in air for 4 hours.

(3) Preparation of MnO_2/g-C_3N_4 composite by ball milling

The solid mixer with a mole ratio of 2 : 3 between manganese acetate tetrahydrate and potassium permanganate was ball milled at a speed of 500 r/min for 5 h. The products were filtered with a large amount of deionized water, and dried at 120 °C. After grinding with g-C_3N_4, the products were heated at 400 °C in air for 4 hours to obtain MnO_2/g-C_3N_4 composite by ball milling, abbreviated as b-MnOCN.

(4) Preparation of MnO$_2$/g-C$_3$N$_4$ composite by microwave

A certain amount of manganese nitrate and g-C$_3$N$_4$ was added to a ceramic boat and heated in the air for 5 min by using microwave irradiation. After natural cooling, the products were calcined at 400 °C for 4 h to prepare MnO$_2$/g-C$_3$N$_4$ composite by microwave, named m-MnOCN.

5.3   Characterization of MnO$_2$/g-C$_3$N$_4$ composite

5.3.1   XRD of MnO$_2$/g-C$_3$N$_4$ composite

Figure 5.1 displays the XRD patterns of p-MnOCN, b-MnOCN, m-MnOCN composite and the pure g-C$_3$N$_4$. The characteristic diffraction peaks at 27.4° and 13.0°, attributed to the (002) and (100) reflections of carbon nitride, could be clearly observed. According to the XRD patterns of the standard samples with JCPDS PDF 72-1982, 41-1442, 80-1098 and 24-0734, the diffraction peaks at about 9.8°, 17.6°, 24.1°, 32.9°, 36.7°, 49.2°, 55.1° of manganese oxides are observed and

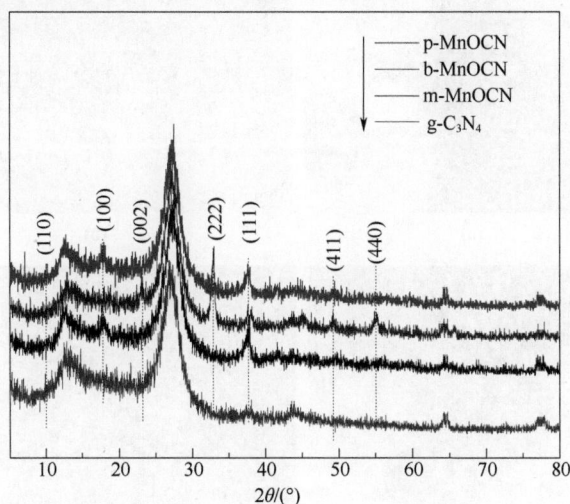

Figure 5.1   XRD patterns of g-C$_3$N$_4$, p-MnOCN, b-MnOCN and m-MnOCN

ascribed to (110), (100), (002), (222), (111), (411) and (440) crystal planes, respectively. In comparison with the g-C$_3$N$_4$ sample, the XRD patterns of MnO$_2$ on g-C$_3$N$_4$ are weaker than those of g-C$_3$N$_4$, suggesting the good dispersion of MnO$_2$ on g-C$_3$N$_4$.

5.3.2 TEM of MnO$_2$/g-C$_3$N$_4$ composite

The morphologies of the prepared MnO$_2$/g-C$_3$N$_4$ photocatalysts are investigated by TEM and HR-TEM, and the results are displayed in Figure 5.2. Figure 5.2(a) and Figure 5.2(b) exhibit the TEM and HR-TEM images of p-MnOCN. Obviously, the MnO$_2$ with a rod-like structure is deposited on the surface of g-C$_3$N$_4$ by the   precipitation method, and it can be seen that the (110) crystal planes of MnO$_2$ with 0.311 nm lattice fringe spacing in the material with p-MnOCN in Figure 5.2(b). The TEM and HR-TEM images of m-MnOCN are shown in Figure 5.2(c) and Figure 5.2(d), and the HR-TEM image reveals that the (111) crystal planes of MnO$_2$ with 0.271 nm lattice fringes spacing of in material with m-MnOCN. It can also be seen that the the (002) crystal planes of MnO$_2$ with 0.326 nm lattice fringes spacing of in material with b-MnOCN in the Figure 5.2(f). Thus, it can be concluded that the manganese oxides with different exposed crystal planes supported on g-C$_3$N$_4$ are successfully prepared by different preparation methods.

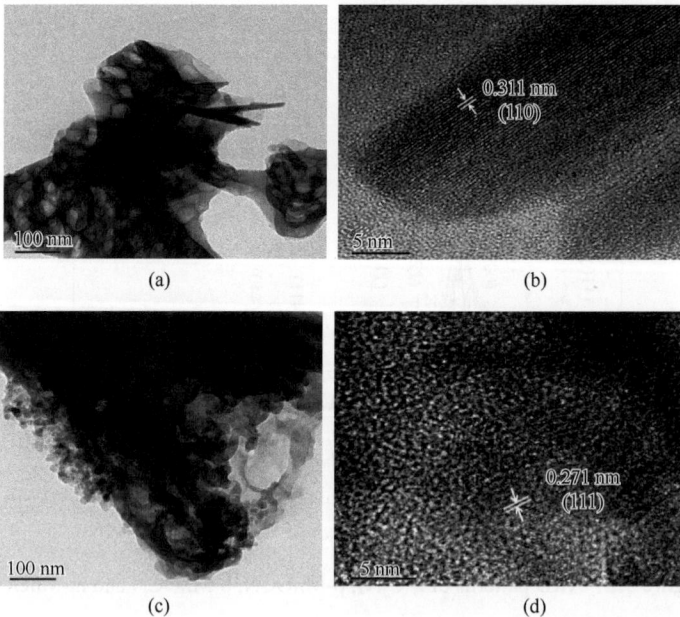

(a)

(b)

(c)

(d)

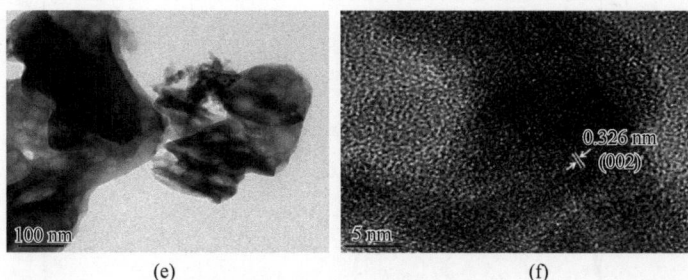

Figure 5.2   TEM and HR-TEM images of (a) and (b) p-MnOCN, (c) and (d)m-MnOCN, (e) and (f) b-MnOCN

5.3.3   UV-vis absorption and PL spectra of MnO$_2$/g-C$_3$N$_4$ composite

It is well acknowledged that the optical and photoelectric performance of photocatalytic materials reasonably determine the photocatalytic activity. UV-Vis absorption can be utilized to evaluate the optical properties of photocatalysts. Figure 5.3 displays the UV-Vis absorption spectra of the synthesized g-C$_3$N$_4$, p-MnOCN, b-MnOCN and m-MnOCN. It can be obviously seen that MnO$_2$/g-C$_3$N$_4$ photocatalyst could significantly increase the absorption range of visible light, which could boost the photocatalytic hydrogen evolution activity. The m-MnOCN with exposed (111) crystal planes exhibits the best visible light absorption among the prepared materials. Figure 5.4 displays the PL spectra for g-C$_3$N$_4$, p-MnOCN, b-MnOCN and m-MnOCN. Generally, weak fluorescence signals are corresponding to the fast separation rate between photogenerated electron and hole, indicating good photocatalytic activity. From the results of Figure 5.4, compared with fresh g-C$_3$N$_4$, the intensity of the peak is significantly weakened in the MnO$_2$/g-C$_3$N$_4$ composite, suggesting that the introduction of MnO$_2$ suppresses the electron-hole pair recombination. The MnO$_2$ with exposed (111) crystal planes displays lower fluorescence signal, suggesting it has better separation rate between photogenerated electron and hole than the other exposed crystal planes. Therefore, the proper introduction of MnO$_2$ with exposed crystal planes is beneficial to upgrade the hydrogen evolution activity.

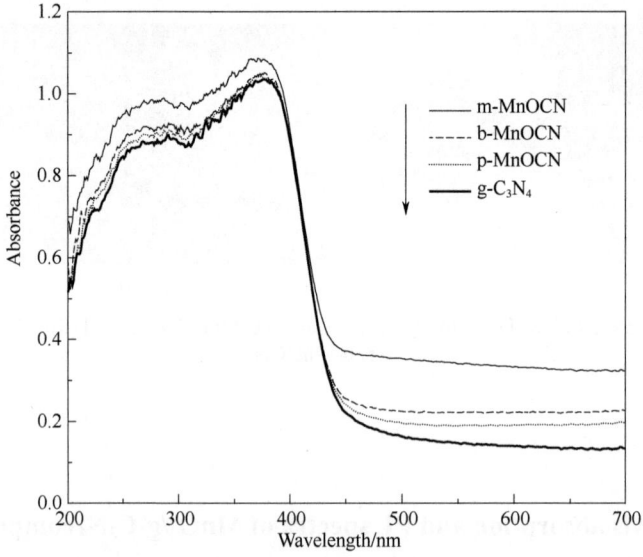

Figure 5.3   UV-vis absorption spectra of the synthesized g-C₃N₄, p-MnOCN, b-MnOCN and m-MnOCN

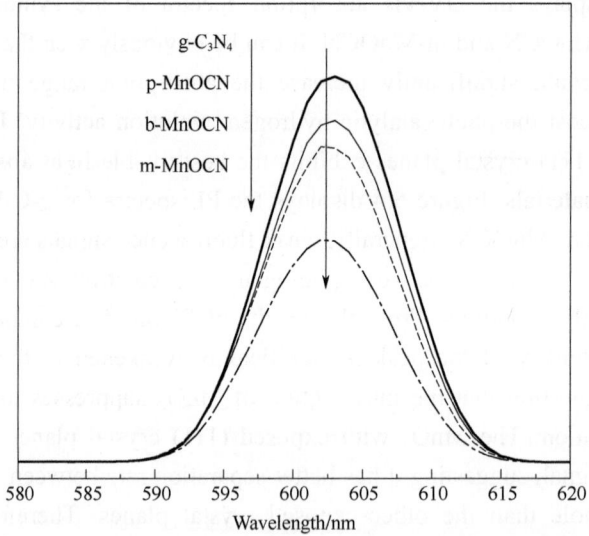

Figure 5.4   PL spectra (λ_{ex} = 370 nm) for the synthesized g-C₃N₄, p-MnOCN, b-MnOCN and m-MnOCN

5.3.4   XPS and EPR of MnO$_2$/g-C$_3$N$_4$ composite

To reveal oxygen vacancies of g-C$_3$N$_4$, p-MnOCN, b-MnOCN and m-MnOCN, EPR characterization was conducted. As shown in Figure 5.5, similar to other g-C$_3$N$_4$ based materials, the prepared samples show a single Lorentzian line centered at 3430 G with the symmetrical EPR signal (g=2.003) originating from the unpaired electrons on the carbon atoms of the π-conjugated aromatic rings of carbon nitride[6]. Notably, m-MnOCN shows the strongest signal, followed by b-MnOCN, p-MnOCN and finally g-C$_3$N$_4$, indicating the order of concentration of lone electron pairs and charge delocalization improvement in the samples, leading to the excellent photocatalytic process with a great density of charge carriers[7].

As the m-MnOCN exhibits the best photocatalytic H$_2$ evolution activity, XPS analysis is further carried out to investigate the detailed chemical compositions of g-C$_3$N$_4$ and m-MnOCN, and the obtained XPS results are displayed in Figure 5.6. In Figure 5.7, three obvious peaks at around 285 eV, 399 eV and 530 eV in both materials could be ascribed to the signals of C 1s, N 1s and O 1s, respectively. Table 5.2 shows the atomic relative content (%) of g-C$_3$N$_4$ and m-MnOCN from XPS characterization. It can be seen that oxygen atomic relative content of m-MnOCN is higher than that of g-C$_3$N$_4$. Figure 5.6(c) presents the XPS spectrum of O 1s. There are two peaks at around 532 eV and 530 eV, corresponding to the oxygen species of hydroxyl group and lattice oxygen, respectively. It can be found that O 1s peaks are much broader and stronger at 532 eV in m-MnOCN than in g-C$_3$N$_4$, meaning that oxygen atoms are obtained on the surface of m-MnOCN. Table 5.1 exhibits the BE and RC of O 1s for the prepared g-C$_3$N$_4$ and m-MnOCN. The relative content of oxygen in the m-MnOCN composite is higher than that in g-C$_3$N$_4$, indicating the rich oxygen content in the prepared m-MnOCN.

Figure 5.6(a) displays the XPS spectra of C 1s of g-C$_3$N$_4$ and m-MnOCN. Three peaks of C 1s spectra in both samples could be found at about 294.1 eV, 287.8 eV, and 284.3 eV[8], corresponding to the signals of π-excitation, the sp^2-hybridized C in N=C—N group and the surface carbon in C-C coordination, respectively[9]. The relative content of π-excitation of m-MnOCN is remarkably higher than that of pure g-C$_3$N$_4$ in Table 5.3. The N 1s spectra of g-C$_3$N$_4$ and m-MnOCN are displayed in Figure 5.6(b). Four peaks at about 398.1 eV, 398.8 eV, 400.3 eV, and 405.0 eV could be deconvoluted to the signals of nitrogen in the C=N—C group[10], tertiary nitrogen bonded to N-C$_3$ group[11], nitrogen in the amino function group[12] and the positive charge effects of nitrogen in heterocycle[13], respectively. The relative content of positive charging effects in heterocycle of the prepared m-MnOCN composite is significantly higher than that of g-C$_3$N$_4$ in Table 5.4. Figure 5.6(d) exhibits the XPS spectra of Mn 2p of m-MnOCN. The observed Mn 2p peaks at 641.3 eV and 653.9 eV are in agreement with Mn 2p$_{3/2}$ and Mn 2p$_{1/2}$, which can be deconvoluted to the coexistence of Mn^{3+} and Mn^{4+} in m-MnOCN. It is noteworthy that defective between Mn^{3+} and Mn^{4+} active sites pave a pathway for fast charge transfer to upgrade the photocatalytic H$_2$ evolution activity from water splitting[14].

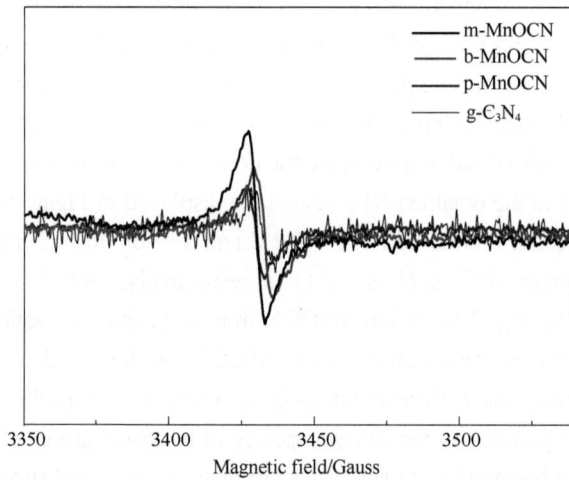

Figure 5.5   EPR spectra for the synthesized g-C$_3$N$_4$, p-MnOCN, b-MnOCN and m-MnOCN

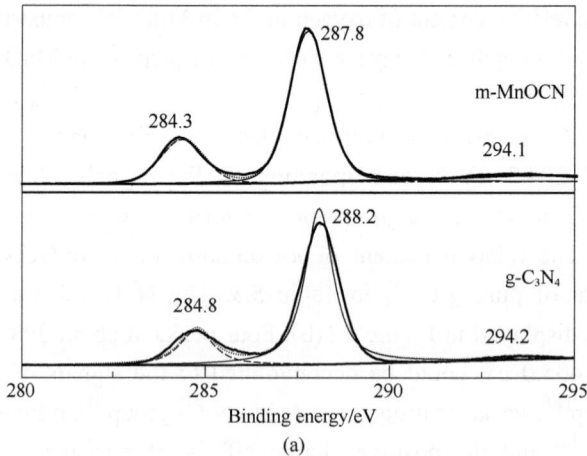

(b)

(c)

(d)

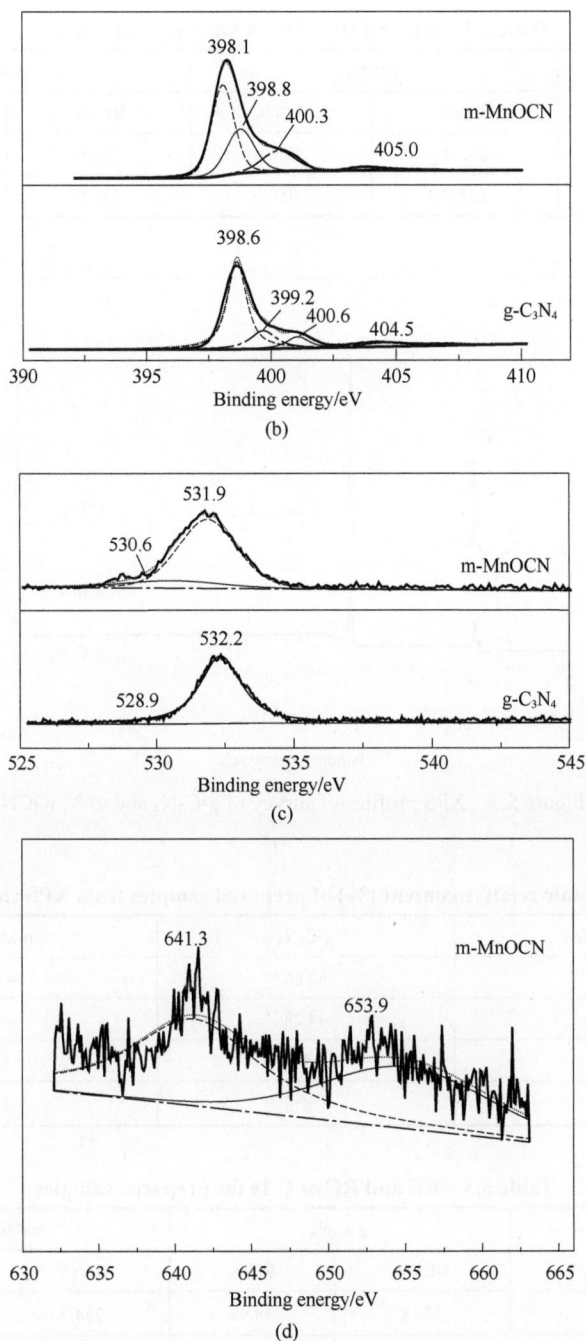

Figure 5.6 XPS profiles of (a) C 1s, (b) N 1s, (c) O 1s and (d) Mn 2p of the prepared g-C$_3$N$_4$ and m-MnOCN

Table 5.1 BE and RC of O 1s for prepared samples

Sample	g-C$_3$N$_4$		m-MnOCN	
	BE/eV	RC/%	BE/eV	RC/%
Lattice oxygen	532.2	99.99	531.9	84.48
Oxygen species	528.9	0.01	530.6	15.51

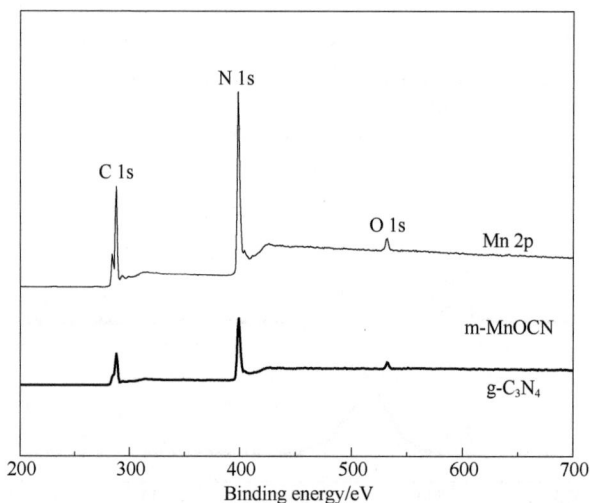

Figure 5.7 XPS profiles of survey of g-C$_3$N$_4$ and m-MnOCN

Table 5.2 Atomic relative content (%) of prepared samples from XPS characterization

Sample	g-C$_3$N$_4$	m-MnOCN
N	52.66	47.63
C	44.28	47.60
O	3.05	3.43
Mn	—	1.34

Table 5.3 BE and RC of C 1s for prepared samples

Sample	g-C$_3$N$_4$		m-MnOCN	
	BE/eV	RC/%	BE/eV	RC/%
C—C	284.8	19.93	284.3	21.93
N=C—N	288.1	74.32	287.8	66.97
π-excitation	293.9	5.75	294.1	11.09

Table 5.4   BE and RC of N 1s for prepared samples

Sample	g-C₃N₄		m-MnOCN	
	BE/eV	RC/%	BE/eV	RC/%
N=C—N	398.6	62.73	398.1	50.20
N doping	399.2	21.10	399.2	29.80
Amino function	400.6	14.84	400.6	16.66
Charging effects	404.5	1.33	405.0	3.34

5.4   Photocatalytic H_2 generation testing MnO_2/g-C_3N_4 composite

5.4.1   Hydrogen production efficiency

Photocatalytic H_2 evolution performance was evaluated with Pt co-catalyst under visible light illumination (λ=420 nm). The pure MnO_2 had poor activity of photocatalytic H_2 production, and the photocatalytic hydrogen evolution of g-C_3N_4, p-MnOCN, b-MnOCN and m-MnOCN rates are shown in Figure 5.8(a). It could be

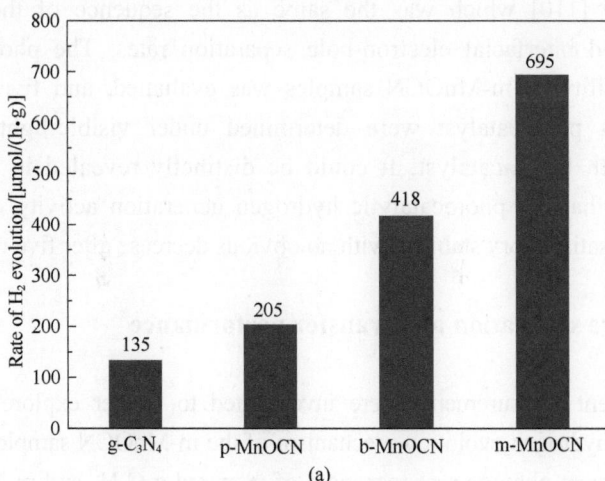

(a)

Figure 5.8

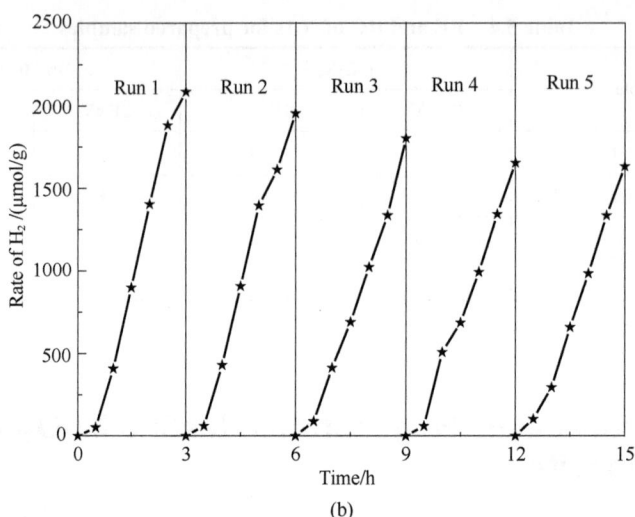

Figure 5.8 (a)The rate of photocatalytic H_2 evolution with 10 vol% TEOA, 1.5 wt% Pt, and 50 mg prepared photocatalysts under visible light (λ = 420 nm), (b) recyclability of m-MnOCN photocatalyst for the photocatalytic H_2 evolution under visible-light irradiation (λ = 420 nm)

obviously seen that the photocatalytic H_2 evolution rates of g-C_3N_4, p-MnOCN, b-MnOCN and m-MnOCN are measured to be 135 μmol/(h·g), 205 μmol/(h·g), 418 μmol/(h·g) and 695 μmol/(h·g), respectively. The photocatalytic performances of p-MnOCN [110], b-MnOCN [002], and m-MnOCN [111] were in the order of [111]> [002]> [110] which was the same as the sequence of the visible-light absorbance and interfacial electron-hole separation rates. The photocatalytic H_2 evolution stability of m-MnOCN samples was evaluated, and five experimental cycles of this photocatalyst were determined under visible-light (λ=420 nm) irradiation with Pt co-catalyst. It could be distinctly revealed in the results of Figure 5.8(b) that the photocatalytic hydrogen generation activity of m-MnOCN still displayed satisfactory stability with no obvious decrease after five recycling runs.

5.4.2 Charge separation and transfer performance

Photocurrent measurements were investigated to further explore the enhanced photocatalytic hydrogen evolution mechanism of the m-MnOCN sample. Figure 5.9(a) shows the transient photocurrent responses of prepared g-C_3N_4 and m-MnOCN. They were performed with an interval light on/off cycle mode by visible-light irradiation.

From the results of Figure 5.9(a), it could be obviously seen that the prepared m-MnOCN composite presents significantly higher photocurrent and more preferable photo-stability than that of g-C$_3$N$_4$, indicating that the charge transfer and the separation process between MnO$_2$ and g-C$_3$N$_4$ are accelerated in the m-MnOCN composite.

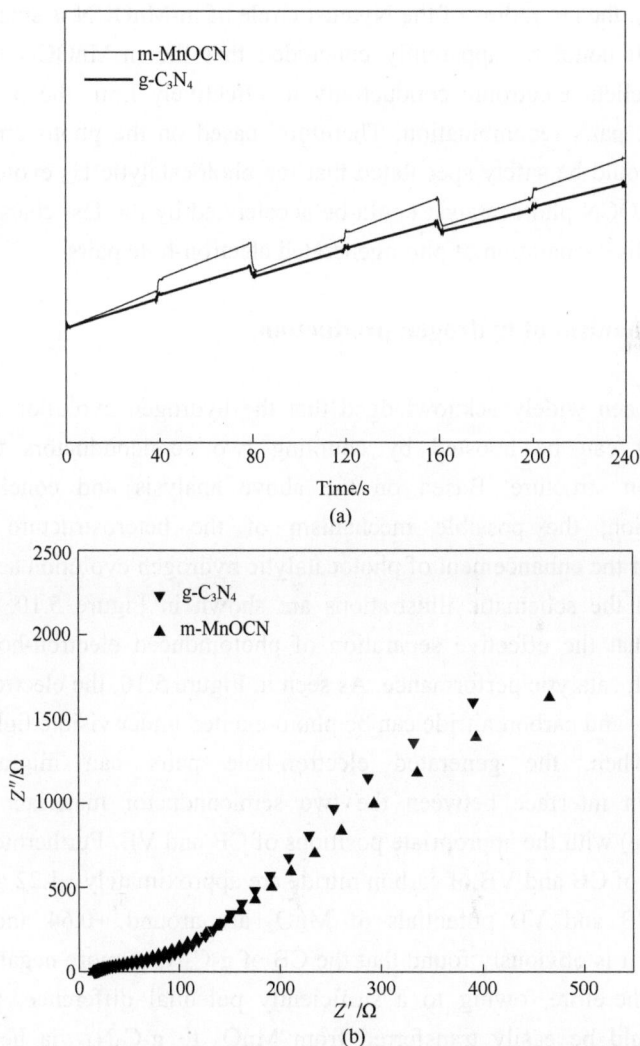

Figure 5.9   (a) Transient photocurrent density of m-MnOCN and g-C$_3$N$_4$ electrodes at 0.3 V versus Ag/AgCl, (b) electrochemical impedance spectra of the as-prepared samples at -0.4 V versus Ag/AgCl

Besides, the EIS is carried out to further identify electron transfer efficiency with the electrodes. Figure 5.9(b) shows the EIS Nyquist plots for m-MnOCN and g-C_3N_4 photocatalysts under visible-light irradiation. It is worthy of notice that the larger arc radius on the EIS Nyquist plot demonstrates the lower effective separation of photogenerated electron-hole pairs as well as better charge transfer ability at the interface across the electrode and electrolyte. As shown in Figure 5.9(b), the arc radius of the Nyquist circle of m-MnOCN is smaller than that of g-C_3N_4. It could be apparently concluded that the m-MnOCN photocatalyst exhibits excellent electronic conductivity to effectively limit the photogenerated electron-hole pairs recombination. Therefore, based on the photocurrent and EIS analysis, it could be safely speculated that the photocatalytic H_2 evolution activity of the m-MnOCN photocatalyst could be accelerated by the fast charge transfer as well as effective separation of photogenerated electron-hole pairs.

5.4.3 Mechanism of hydrogen production

It has been widely acknowledged that the hydrogen evolution activity of a photocatalyst can be boosted by coupling two semiconductors to form the heterojunction structure. Based on the above analysis and conclusions from characterization, the possible mechanism of the heterostructure m-MnOCN composite for the enhancement of photocatalytic hydrogen evolution activity can be inferred, and the schematic illustrations are shown in Figure 5.10. It is widely recognized that the effective separation of photoinduced electron-hole pairs can improve photocatalytic performance. As seen in Figure 5.10, the electron-hole pairs in both MnO_2 and carbon nitride can be photo-excited under visible light irradiation condition. Then, the generated electron-hole pairs can migrate via the heterojunction interface between the two semiconductor materials (MnO_2 and carbon nitride) with the appropriate positions of CB and VB. Furthermore, the band gap energies of CB and VB of carbon nitride are approximately -1.22 and $+1.56$ eV, while the CB and VB potentials of MnO_2 are around $+0.64$ and $+2.28$ eV, respectively. It is obviously found that the CB of g-C_3N_4 is more negative than that of MnO_2. Therefore, owing to a sufficiently potential difference, photoexcited electrons could be easily transferred from MnO_2 to g-C_3N_4 via heterostructure interface, which declines the possibility of electron-hole recombination. Then more electrons can be generated on the g-C_3N_4 surfaces to further upgrade the hydrogen evolution activity of photocatalyst. The heterojunction structure of MnO_2/g-C_3N_4

composite paves the way for dropping the interface energy and expediting the electron transfer process. The Z scheme mechanism is proposed for the photocatalytic hydrogen production mechanism of the MnO_2/g-C_3N_4 composite. From the mechanism of Figure 5.10, the photo-generated electrons on the CB of MnO_2 can be migrated with the VB of g-C_3N_4 by the potential difference, therefore, separation of electrons and holes could be greatly accelerated. The electrons accumulated on the CB of g-C_3N_4 that are further transferred to the Pt nanoparticles and the reduction of protons to H_2 could be catalyzed, while the holes in VB of MnO_2 are consumed by TEOA. Besides, the MnO_2/g-C_3N_4 composite could greatly limit the migration of photogenerated carriers on g-C_3N_4 to boost the water splitting efficiency. It can been concluded that the hydrogen production efficiency of the photocatalyst will be enhanced due to the effective separation between electrons and holes.

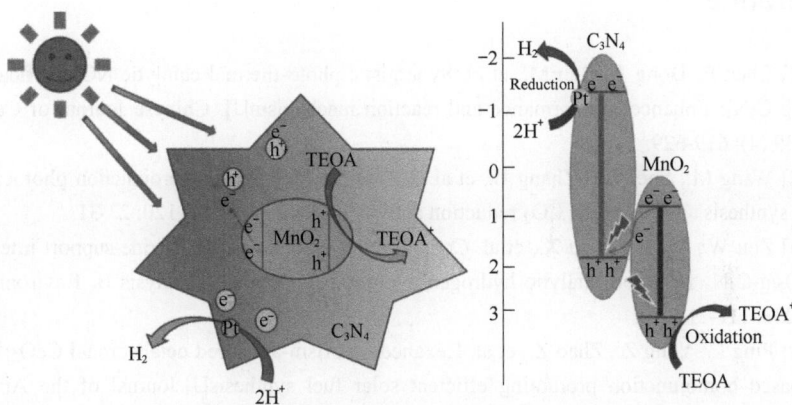

Figure 5.10   Schematic illustrations of the proposed photocatalytic mechanism over the MnO_2/g-C_3N_4 composite under visible light irradiation

5.5 Conclusion

The manganese dioxides with different exposed crystal planes supported on g-C_3N_4 were prepared successfully and examined by a series of characterizations to confirm the physical and chemical properties. The prepared composites were evaluated to obtain the performance of photocatalytic hydrogen generation under visible-light irradiation. The photocatalytic performances of manganese dioxides with different exposed crystal planes (p-MnOCN [110], b-MnOCN [002], and m-MnOCN [111]) were in the order of [111]>[002]>[110] which was the same as the sequence of the visible-light absorbance and interfacial electron-hole separation rates. The prepared m-MnOCN composite exhibits the best photocatalytic capacity for H_2 evolution rate of 695 μmol/(h·g) under visible light (λ= 420 nm) illumination. The photocatalytic hydrogen generation capacity of m-MnOCN still presents satisfactory stability after four recycling runs. Schematic illustrations of mechanism indicate that Z-scheme mechanism could be the possible mechanism for photocatalytic H_2 production reactions. This study displays that engineering the exposed crystal-plane-dependent metal oxide on the semiconductors could provide a scientific basis for the reasonable design of efficient photocatalysts.

Reference

[1] Chen P., Dong F., Ran M., et al. Synergistic photo-thermal catalytic NO purification of MnO_x/g-C_3N_4: Enhanced performance and reaction mechanism[J]. Chinese Journal of Catalysis 2018, 39 (4): 619-629.

[2] Wang M., Shen M., Zhang L., et al. 2D-2D MnO_2/g-C_3N_4 heterojunction photocatalyst: In-situ synthesis and enhanced CO_2 reduction activity[J]. Carbon ,2017, 120: 23-31.

[3] Zou W., Deng B., Hu X., et al. Crystal-plane-dependent metal oxide-support interaction in CeO_2/g-C_3N_4 for photocatalytic hydrogen evolution[J]. Applied Catalysis B: Environmental, 2018, 238: 111-118.

[4] Ping L., Yong Z., Zhao Z., et al. Hexahedron prism-anchored octahedronal CeO_2: crystal facet-based homojunction promoting efficient solar fuel synthesis[J].Journal of the American Chemical Society ,2015, 137 (30): 9547-9550.

[5] Xia P., Zhu B., Cheng B., et al. 2D/2D g-C_3N_4/MnO_2 nanocomposite as a direct Z-scheme photocatalyst for enhanced photocatalytic activity[J]. ACS Sustainable Chemistry & Engineering, 2018, 6 (1): 965-973.

[6] Zeng D., Xu W., Ong W. J., et al. Toward noble-metal-free visible-light-driven photocatalytic hydrogen evolution: monodisperse sub-15nm Ni_2P nanoparticles anchored on porous g-C_3N_4 nanosheets to engineer 0D-2D heterojunction interfaces[J]. Applied Catalysis B: Environmental, 2018, 221: 47-55.

[7] Zhang Y. C., Li Z., Zhang L., et al. Role of oxygen vacancies in photocatalytic water oxidation on ceria oxide: Experiment and DFT studies[J]. Applied Catalysis B: Environmental, 2018, 224: 101-108.

[8] Xu H., Xiao R., Huang J.,et al. In situ construction of protonated g-C$_3$N$_4$/Ti$_3$C$_2$ MXene Schottky heterojunctions for efficient photocatalytic hydrogen production[J].Chinese Journal of Catalysis, 2021, 42 (1):107-114.

[9] Liu X., Zhang Q., Liang L., et al. In-situ growing of CoO nanoparticles on g-C$_3$N$_4$ composites with highly improved photocatalytic activity for hydrogen evolution[J]. Royal Society Open Science, 2019, 6 (7): 190433.

[10] Gao D., Xu Q., Zhang J., et al. Defect-related ferromagnetism in ultrathin metal-free g-C$_3$N$_4$ nanosheets[J]. Nanoscale, 2014, 6 (5): 2577-2581.

[11] Liu X., He L., Chen X., et al. Facile synthesis of CeO$_2$/g-C$_3$N$_4$ nanocomposites with significantly improved visible-light photocatalytic activity for hydrogen evolution[J]. International Journal of Hydrogen Energy, 2019, 44 (31): 16154-16163.

[12] Shen R., He K., Zhang A., et al. In-situ construction of metallic Ni$_3$C@Ni core-shell cocatalysts over g-C$_3$N$_4$ nanosheets for shell-thickness-dependent photocatalytic H$_2$ production[J]. Applied Catalysis B: Environmental, 2021, 291: 120104.

[13] Wang S., Li C., Wang T., et al. Controllable synthesis of nanotube-type graphitic C$_3$N$_4$ and their visible-light photocatalytic and fluorescent properties[J].Journal of Materials Chemistry A, 2014, 2 (9): 2885-2890.

[14] Mo Z., Xu H., Chen Z., et al. Construction of MnO$_2$/monolayer g-C$_3$N$_4$ with Mn vacancies for Z-scheme overall water splitting[J]. Applied Catalysis B: Environmental, 2019, 241: 452-460.

Chapter 6

LaVO$_4$/g-C$_3$N$_4$ composite with oxygen defect

6.1 Background

Heterojunction-structure g-C$_3$N$_4$ photocatalysts have been designed to elevate the photocatalytic activity of hydrogen evolution. Song et al.[1] prepared a Z-scheme BiVO$_4$/g-C$_3$N$_4$ nanocomposite to realize robust photocatalytic hydrogen evolution. The Z-scheme mechanism between BiVO$_4$ and g-C$_3$N$_4$ is in favor of enhanced photocatalytic H$_2$ evolution activity. Luo et al.[2] synthesized g-C$_3$N$_4$/SrTiO$_3$ nanocomposite and it exhibits more visible light absorption and faster photo-generated charge transfer to promote water splitting activity to evoluate H$_2$. The heterojunction-structure g-C$_3$N$_4$ nanocomposite show better photocatalytic performance than that of fresh g-C$_3$N$_4$, because of high visible-light absorbance and separation efficiency of photogenerated electron-hole pairs, as well as the fast charge transportation[3]. However, it remains a great challenge to improve the oxygen evolution over these g-C$_3$N$_4$ based materials.

Vanadate has been assumed as one of the most promising semiconductors for photocatalytic water oxidation owing to its moderate energy gap and in favor of band alignment. Lanthanum vanadate (LaVO$_4$) is considered to be an excellent photocatalytic material with immense potential, because its V 3d orbital electrons can be activated by visible light, which can be facilely formed stable monoclinic structure with high coordination number[4]. It is also reported that the activity of the LaVO$_4$ photocatalyst can be significantly improved by forming a heterojunction structure and introducing defects. Wang et al. [5] prepared defect-rich carbon nitride through thermal defects, which effectively inhibited charge recombination and broadened light absorption, thereby improving the activity of the photocatalyst. Wang et al.[6] proposed an oxygen defect-mediated Z-scheme mechanism for charge separation in heterojunctions. The prepared g-C$_3$N$_4$/OD-ZnO photocatalyst shows higher hydrogen evolution activity based on visible light Z-scheme, and the optimal hydrogen evolution rate is about 5 times that of pure g-C$_3$N$_4$. La$_2$O$_3$ is used as a co-catalyst for enhanced photocatalytic O$_2$ production from water oxidation over g-C$_3$N$_4$ Z-scheme photocatalysts[7]. The reported papers on the photocatalysts for water splitting over LaVO$_4$ material are rare.

Thus, in this book, the LaVO$_4$/g-C$_3$N$_4$ composite is simply prepared and the hydrogen and oxygen evolution activity of the prepared photocatalyst was evaluated by the photocatalytic water splitting test. Meanwhile, through a series of characterizations, the chemical composition, morphology, optical and photoelectrochemical properties were examined. Finally, the enhanced mechanism of photocatalytic activity for water splitting was discussed.

6.2   Preparation of LaVO$_4$/g-C$_3$N$_4$ composite

Urea, ammonium vanadate, lanthanum oxide, nitric acid and triethanolamine (TEOA) were purchased from Aladdin Industrial Corporation (Shanghai, China). These chemicals were used without further processing.

(1) LaVO$_4$ preparation

Firstly, 6 mol/L HNO$_3$ aqueous solution was used to dissolve a certain mass of lanthanum oxide. Then, the obtained lanthanum nitrate solution was mixed with 0.4 mol/L ammonium vanadate aqueous solution and transferred into a 100 mL autoclave with Teflon-lined stainless steel. The solution in the autoclave was kept at a hydrothermal temperature of 180 °C for 24 h. After the hydrothermal process, the precipitates were filtered and washed with distilled water and dried at 80 °C in a drying cabinet to obtain LaVO$_4$.

(2) LaVO$_4$/g-C$_3$N$_4$ composite preparation

Firstly, the fresh g-C$_3$N$_4$ material was prepared by calcination of urea at 500 °C for 3 h with the heating speed of 10 °C/min in the covered crucible. Then, 200 mg g-C$_3$N$_4$ powder and a certain mass of LaVO$_4$ were put into the flask with 50 mL of deionized water and kept stirring for 20 h in the 70 °C water bath. After rotavaporating, the LaVO$_4$/g-C$_3$N$_4$ composite (0.5%~10% mass ratio of LaVO$_4$ to C$_3$N$_4$) was obtained by drying at 80 °C. The prepared materials were denoted as LVOCN$_x$, where x refers to the mass ratio of LaVO$_4$ to C$_3$N$_4$ (x = 0.5, 1, 5 and 10), which could be abbreviated as LVOCN0.5, LVOCN1, LVOCN5 and LVOCN10. The LaVO$_4$/g-C$_3$N$_4$ composite with O defects was prepared by the calcination at 300 °C, abbreviated as LVOCN1-O.

6.3 Characterization of LaVO₄/g-C₃N₄ composite

6.3.1 XRD of LaVO₄/g-C₃N₄ composite

Figure 6.1 (a) shows the XRD patterns of the pure g-C₃N₄ and nitrogen-rich LaVO₄/g-C₃N₄ composite. It can be clearly observed that the tharacteristic diffraction peaks at 27.4° and 13.0° are attributed to the (002) and (100) reflections of carbon nitride. These two apparent characteristic peaks appearing in all catalysts indicate that the introduction of LaVO₄ does not change the structure of graphitic carbon nitride. As shown in Figure 6.1 (a), the XRD patterns of fresh LaVO₄ exhibit the monoclinic structure according to the JCPDS No. 500367. Figure 6.1 (b) exhibits the XRD patterns of LaVOCN1 and LaVOCN1-300. As displayed in Figure 6.1 (b), there is no phase change in oxygen-defective LaVOCN1-300 compared with the LaVOCN1, and no peak of any other impurity is detected. Besides, the characteristic diffraction peaks of LaVO₄/g-C₃N₄ composite demonstrated a coexistence of both LaVO₄ and g-C₃N₄ phases. With the increase in LaVO₄ concentration, the LaVO₄ diffraction peak becomes more apparent and sharp, which can be explained by the raise of LaVO₄ grain size.

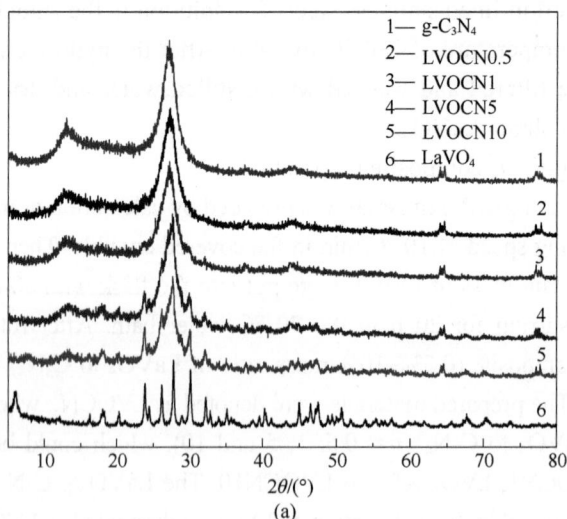

1— g-C₃N₄
2— LVOCN0.5
3— LVOCN1
4— LVOCN5
5— LVOCN10
6— LaVO₄

2θ/(°)

(a)

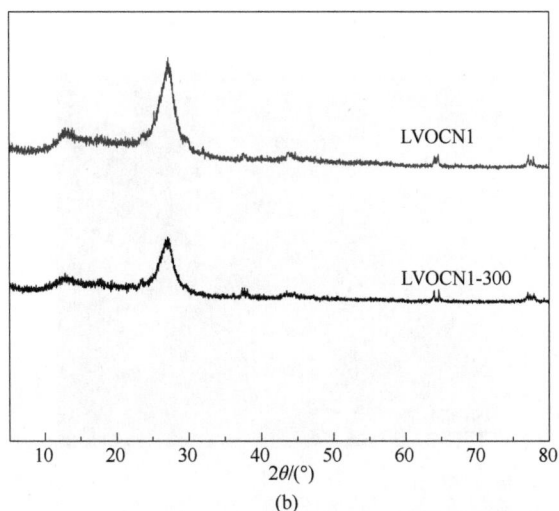

(b)

Figure 6.1   (a) XRD patterns of g-C$_3$N$_4$, LaVOCN0.5, LaVOCN1, LaVOCN5, LaVOCN10 and LaVO$_4$, (b) XRD patterns of LaVOCN1 and LaVOCN1-300

6.3.2   TEM of LaVO$_4$/g-C$_3$N$_4$ composite

The morphologies of the prepared photocatalysts are examined by TEM, and the obtained results are appeared in Figure 6.2. Figure 6.2(a) and Figure 6.2(b) exhibit the TEM images of fresh g-C$_3$N$_4$ and LaVO$_4$, respectively. Obviously, the as-prepared fresh g-C$_3$N$_4$ has nearly transparent characteristics, which indicates that fresh g-C$_3$N$_4$ has an ultrathin thickness with a multilayer structure. From the TEM results of LVOCN1 composite in Figure 6.2(c) and Figure 6.2(d), it can be seen that the multi-layer nanosheet structure is retained. However, compared with fresh g-C$_3$N$_4$, the surface of nanosheet becomes fluffy and rough with soft and loose aggregates. The BET surface areas of composites are slightly higher than those of g-C$_3$N$_4$, as shown in Figure 6.3. The HR-TEM image in Figure 6.2(d) further reveals the (200) crystal planes of LaVO$_4$ with 0.38 nm lattice fringes spacing of in material with LaVOCN1-300. It is also found that LaVO$_4$ nanoparticles are formed with oxygen defect by calcination. Besides, LaVO$_4$ nanoparticles are highly dispersed on the g-C$_3$N$_4$ sheets, which can be inferred that LaVO$_4$/g-C$_3$N$_4$ with heterojunction nanosheets is prepared successfully.

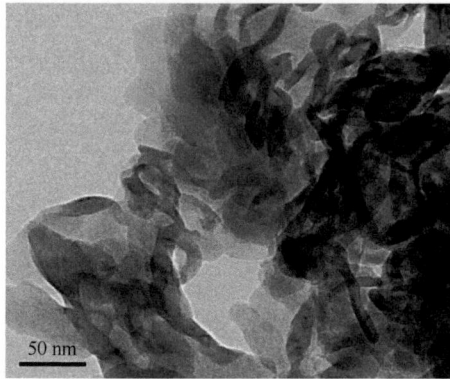

(a)

(b)

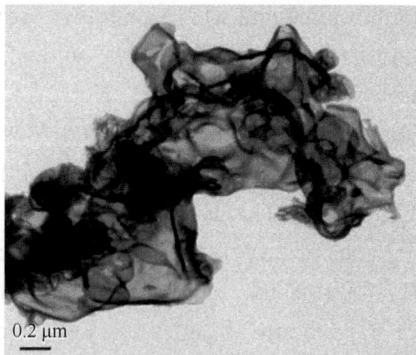

(c)

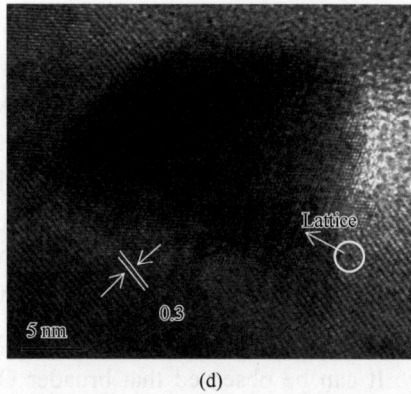

(d)

Figure 6.2   TEM and HR-TEM images of (a) g-C₃N₄, (b) LaVO₄, (c) and (d) LaVOCN1-300

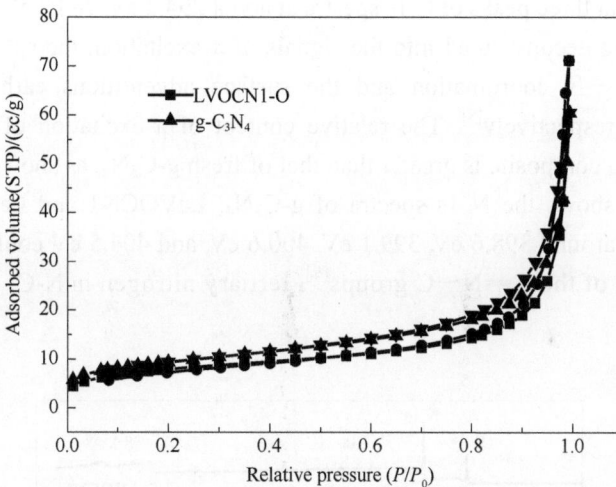

Figure 6.3   The N₂ adsorption/desorption isotherms of g-C₃N₄ and LVOCN1-O

6.3.3   XPS of LaVO₄/g-C₃N₄ composite

In order to obtain detailed chemical compositions of the prepared samples, XPS analysis of g-C₃N₄, LaVOCN1, LaVOCN1-300 and LaVO₄ is further carried out. Figure 6.4 exhibits the obtained XPS results. In Figure 6.4(a), three distinct peaks at about 285 eV, 399 eV and 530 eV in the prepared materials could be regarded as the signals of C 1s, N 1s and O 1s respectively. Table 6.1 exhibits the BE and RC of O 1s for the prepared g-C$_3$N$_4$, LaVOCN1 and LaVOCN1-300. The relative content of oxygen defects of LaVOCN1-300 composite is bigger than that of LaVOCN1, which determines the rich oxygen defect in the prepared LaVOCN1-300 composite. Table 6.2 displays atomic relative content (%) of prepared samples from XPS characterization. It can be seen that oxygen atomic relative content of LaVOCN1-300 is lower than that of LaVOCN1. Figure 6.4(d) shows the XPS spectrum of O 1s. Two peaks at about 529 eV and 532 eV could be ascribed to the lattice oxygen and oxygen species of hydroxyl group adsorbed on the surface, respectively. It can be observed that broader O 1s peaks with strong shoulder at 532 eV in LaVOCN1 and LaVOCN1-300, meaning that oxygen defects are introduced into the surface of LaVO$_4$/g-C$_3$N$_4$ samples successfully. Figure 6.4(b) presents the XPS spectra of C 1s of g-C$_3$N$_4$, LaVOCN1 and LaVOCN1-300. It can be observed that three peaks of C 1s spectra at about 294.2 eV, 288.2 eV, and 284.8 eV, which could be deconvoluted into the signals of π-excitation, the sp^2-hybridized C in the N=C—N coordination and the surface adventitious carbon in C—C coordination, respectively[8]. The relative content of π-excitation of the prepared LaVO$_4$/g-C$_3$N$_4$ composite is greater than that of fresh g-C$_3$N$_4$, as shown in Table 6.3. Figure 6.4(c) shows the N 1s spectra of g-C$_3$N$_4$, LaVOCN1 and LaVOCN1-300. Four peaks at around 398.6 eV, 399.1 eV, 400.6 eV, and 404.5 eV could be ascribed to the signals of the C=N—C groups[9], tertiary nitrogen in N-C$_3$ groups[10], the

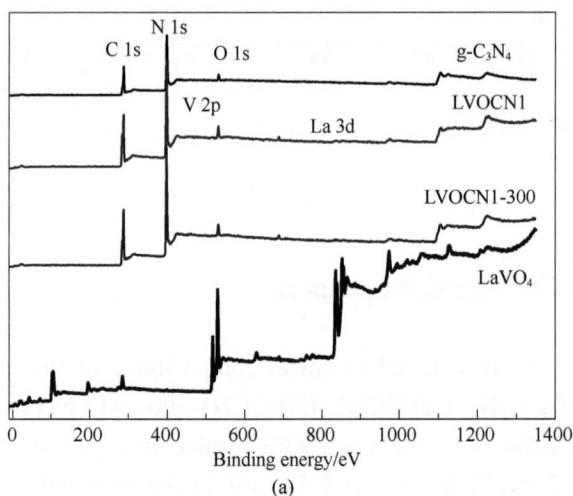

(a)

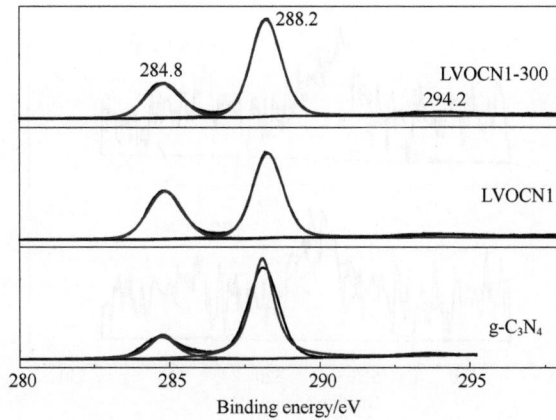

(b)

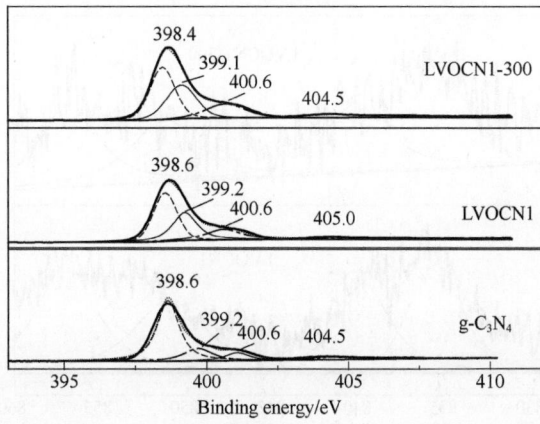

(c)

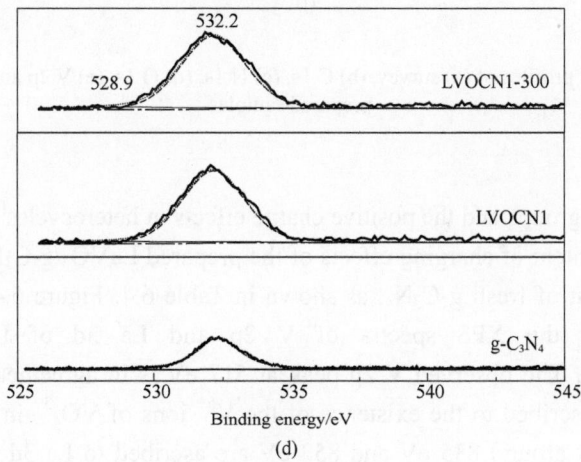

(d)

Figure 6.4

(e)

(f)

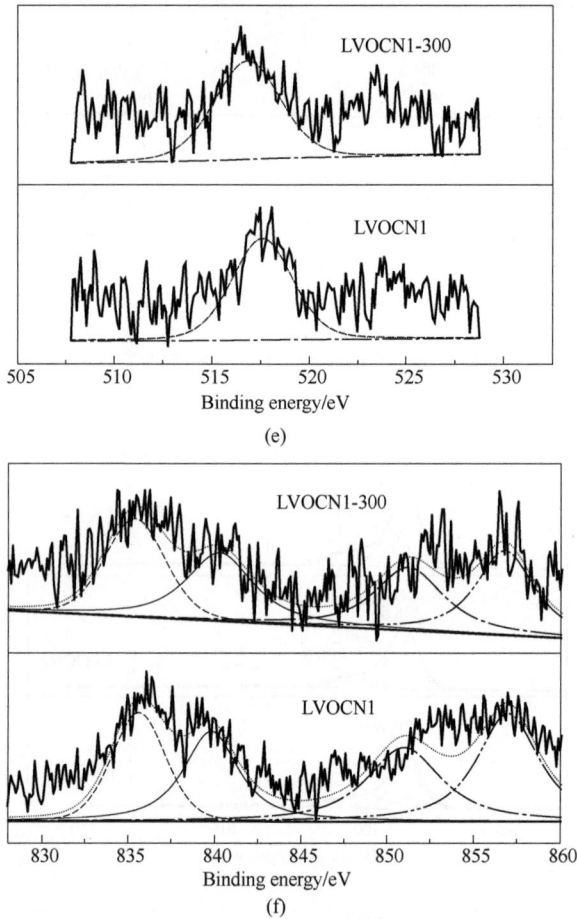

Figure 6.4   XPS profiles of (a) survey, (b) C 1s, (c) N 1s, (d) O 1s, (e) V 2p and (f) La 3d of the prepared samples

amino function groups and the positive charge effects in heterocycles[11], respectively. The relative content of charging effects of the prepared $LaVO_4/g$-C_3N_4 composite is greater than that of fresh g-C_3N_4, as shown in Table 6.4. Figure 6.4(e) and Figure 6.4(f) presents the XPS spectra of V 2p and La 3d of LaVOCN1 and LaVOCN1-300. The observed V 2p peak at 517 eV is in agreement with V 2p$_{3/2}$, which can be ascribed to the existence of the V^{5+} ions of VO_4^{3-} in LaVO$_4$[12].Two typical peaks at around 835 eV and 852 eV are ascribed to La 3d$_{5/2}$ and La 3d$_{3/2}$, respectively, which confirm the existence of the typical La^{3+} in LaVO$_4$[13].

Table 6.1   BE and RC of O 1s for prepared samples

Sample	g-C₃N₄		LVOCN1		LVOCN1-O	
	BE/eV	RC/%	BE/eV	RC/%	BE/eV	RC/%
Lattice oxygen	532.2	99.99	532.2	87.31	532.2	92.08
Oxygen species	528.9	0.01	528.9	12.69	528.9	7.92

Table 6.2   Atomic relative content (%) of prepared samples from XPS characterization

Sample	g-C₃N₄	LaVOCN1	LaVOCN1-300
N	52.66	43.93	44.34
C	44.28	51.64	51.02
O	3.05	4.37	4.12
La	—	0.01	0.01
V	—	0.05	0.51

Table 6.3   BE and RC of C 1s for prepared samples

Sample	g-C₃N₄		LVOCN1		LVOCN1-300	
	BE/eV	RC/%	BE/eV	RC/%	BE/eV	RC%
C—C	284.8	19.93	284.8	36.35	284.8	27.50
N=C—N	288.1	74.32	288.3	57.83	288.2	66.66
π-excitation	293.9	5.75	294.5	5.82	294.2	5.84

Table 6.4   BE and RC of N 1s for prepared samples

Sample	g-C₃N₄		LVOCN1		LVOCN1-300	
	BE/eV	RC/%	BE/eV	RC/%	BE/eV	RC%
N=C—N	398.6	62.73	398.6	43.30	398.4	41.02
N doping	399.2	21.10	399.2	32.88	399.1	32.11
Amino function	400.6	14.84	400.6	18.78	400.6	17.68
Charging effects	404.5	1.33	405.0	5.04	404.5	9.18

6.3.4   UV-vis absorption and PL spectra of LaVO₄/g-C₃N₄ composite

It is known to all that the optical and photoelectric properties of photocatalysts considerably determine its photocatalytic activity. Photoluminescence spectra and

UV-Vis absorption are usually used to evaluate the optical properties of catalysts. Figure 6.5 (a) exhibits the UV-Vis absorption spectra of carbon nitride, LVOCN0.5, LVOCN1, LVOCN5, LVOCN10 and LVOCN15. It can be clearly seen that $LaVO_4/g-C_3N_4$ material can significantly increase the visible light absorption range, which will enhance the photocatalytic hydrogen evolution activity. Figure 6.5 (b)

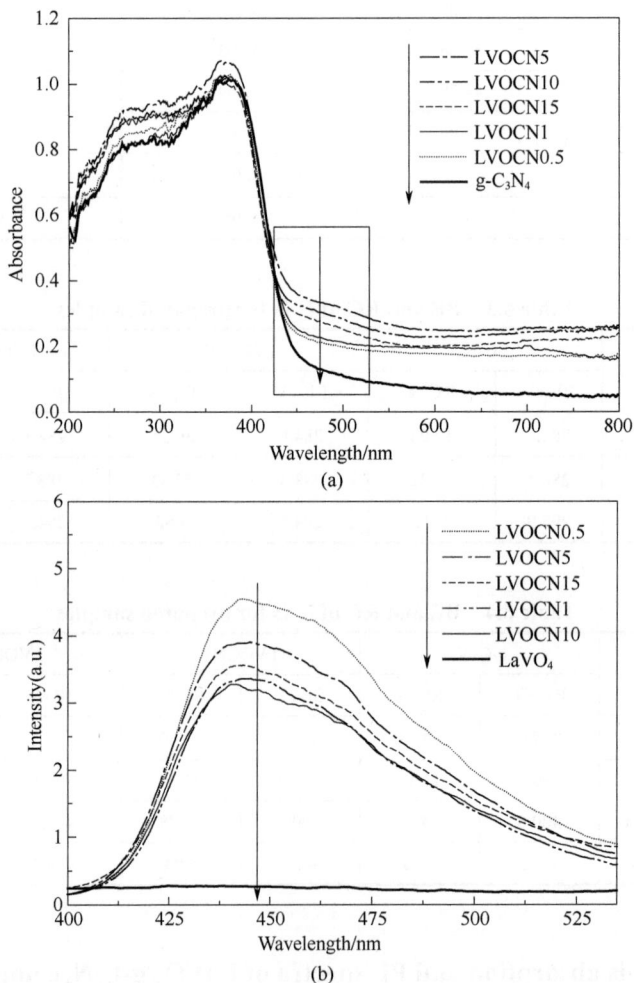

Figure 6.5   (a) UV-vis absorption spectra of the synthesized $g-C_3N_4$, LVOCN0.5, LVOCN1, LVOCN5, LVOCN10 and LVOCN15, (b) PL spectra ($\lambda_{ex} = 370$ nm) for the synthesized $LaVO_4$, LVOCN0.5, LVOCN1, LVOCN5, LVOCN10 and LVOCN15 nanocomposite

displays the PL spectra for LaVO$_4$, LVOCN0.5, LVOCN1, LVOCN5, LVOCN10 and LVOCN15 composite. Generally speaking, strong fluorescence signals indicate that there is more recombination probability between photogenerated electron and hole, which will result in lower photocatalytic activity. From the results of Figure 6.5 (b), when the mass ratio of LaVO$_4$ gradually increases, the intensity of the peak is significantly weakened, suggesting that the introduction of LaVO$_4$ suppresses the electron-hole pair recombination. Meanwhile, the weaker intensity of PL emission signifies the faster photoelectron transfer. Therefore, it can be concluded that the proper introduction of LaVO$_4$ is beneficial to improve the hydrogen evolution performance of the photocatalyst.

6.4   Photocatalytic water splitting testing

6.4.1   Water splitting efficiency

Photocatalytic water-splitting studies were measured by an on-line analysis system (CEL-SPH2N, AG, CEAULIGHT, Beijing) in the reaction vessel with top-irradiation. For H$_2$ production, 50 mg of photocatalytic material was added into 50mL aqueous solution contained TEOA and water. 1.5 wt% of Pt nanoparticles were in-situ doped on the surface of the samples by photodeposition process from the precursor H$_2$PtCl$_6 \cdot$6H$_2$O. For O$_2$ production, 50 mg of photocatalyst was added into 50mL 0.005 mol/L AgNO$_3$ aqueous solution. The reaction vessel temperature was kept at about 10°C by a flow of cooling water. Before the irradiation experiments, the reaction system was evacuated for 30 min to remove air. The visible light source was the 300 W xenon lamp with a 400 nm cut-off filter. The amount of generated hydrogen was determined by gas chromatography (GC 7920, Beijing) by using nitrogen as the carrier gas.

Photocatalytic water splitting performance of carbon nitride, LVOCN0.5, LVOCN1, LVOCN1-O, LVOCN5, LVOCN10 and LaVO$_4$ are examined under visible-light illumination ($\lambda \geqslant 400$ nm) with co-catalyst. The photocatalytic H$_2$ evolution rates are shown in Figure 6.6(a). Obviously, it can be observed that with the increase of introduced LaVO$_4$, the photocatalytic water splitting performance of the composite materials shows a tendency to decrease after increasing. LVOCN1-O exhibits excellent photocatalytic H$_2$ [0.89 mmol/(h·g)] and O$_2$ [0.23 mmol/(h·g)] evolution rate, about 27% and 35% higher than that of LVOCN1, respectively, which can be manifested that photocatalytic water splitting performance could be enhanced by the introduction of appropriate $LaVO_4$ and the oxygen defects. The

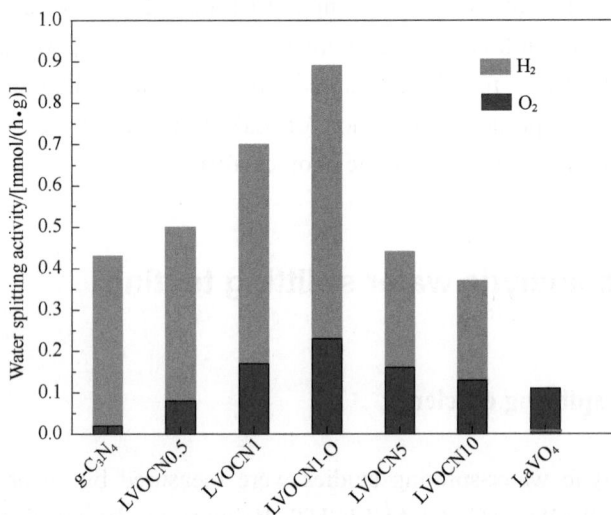

(a)

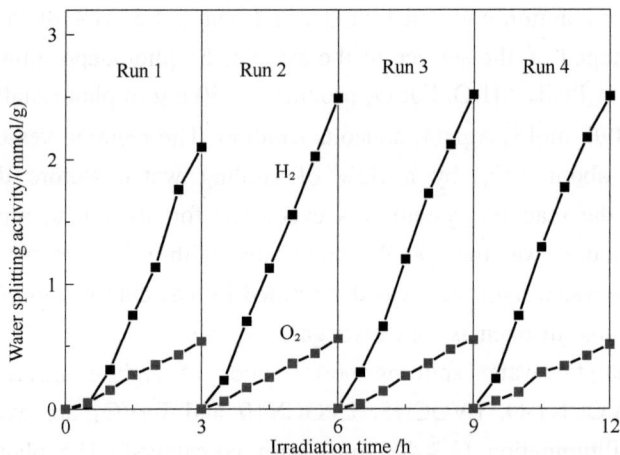

(b)

Figure 6.6 (a) The rate of photocatalytic H_2 and O_2 evolution, (b) recyclability of photocatalyst for photocatalytic water splitting activity under visible-light irradiation ($\lambda \geqslant 400$ nm). Condition: H_2 evolution with 10 vol% TEOA, 1.5 wt% Pt, O_2 evolution with 0.005mol/L $AgNO_3$ aqueous solution

photocatalytic water splitting stability of the prepared sample is determined, and four cycling experiments of the catalyst were measured under visible light ($\lambda \geqslant$ 400 nm) with co-catalyst. From the results of Figure 6.6(b), it is distinctly revealed that the photocatalytic hydrogen and oxygen generation activity of the photocatalyst still displays excellent stability after four recycling runs. It can be reasonably inferred that the prepared oxygen-defect LaVO$_4$/g-C$_3$N$_4$ composite with heterojunction structure indeed shows prominent photocatalytic water splitting activity and stability, which can be explained by the fact that the introduction of appropriate LaVO$_4$ and oxygen defects can effectively improve the transport efficiency of photogenerated carriers by hindering the recombination rate of photogenerated electron-hole pairs.

6.4.2   Charge separation and transfer performance

Photocurrent measurements were carried out to further examine the enhanced photocatalytic mechanism of LaVO$_4$/g-C$_3$N$_4$ photocatalyst. Figure 6.7(a) shows that transient photocurrent responses of prepared LaVO$_4$, LVOCN1 and LVOCN1-O have been performed and compared with an interval light on/off cycle mode by visible-light illumination. From the results of Figure 6.7(a), it has become apparent that the prepared LVOCN1-O composite displays higher photocurrent and preferable photo-stability than that of LVOCN1, indicating the charge transfer and the separation process between LaVO$_4$ and carbon nitride are enhanced in the LVOCN1-O composite.

In addition, the EIS is measured to identify electron transfer efficiency at the electrodes. Figure 6.7(b) shows the EIS Nyquist plots for LaVO$_4$, LVOCN1, and LVOCN1-O catalysts under visible-light irradiation. It is well known, the smaller arc radius on the EIS Nyquist plot demonstrates the higher effective separation of photogenerated electron-hole pairs, as well as the interface charge transfer ability across the electrode and electrolyte would be better. As shown in Figure 6.7(b), the radius of the Nyquist circle for LVOCN1 and LVOCN1-O is smaller than LaVO$_4$, and that of LVOCN1 is smaller than that of LVOCN1-O. It was obviously concluded that the LVOCN1 composite exhibits a stronger electronic conductivity to effectively separate the photogenerated electron-hole pairs. Therefore, we can safely speculate that, based on the photocurrent and EIS analysis, the photocatalytic H$_2$ evolution activity of oxygen-defect LaVO$_4$/g-C$_3$N$_4$ photocatalyst could be promoted by the fast charge transfer and effective separation of photogenerated electron-hole pairs.

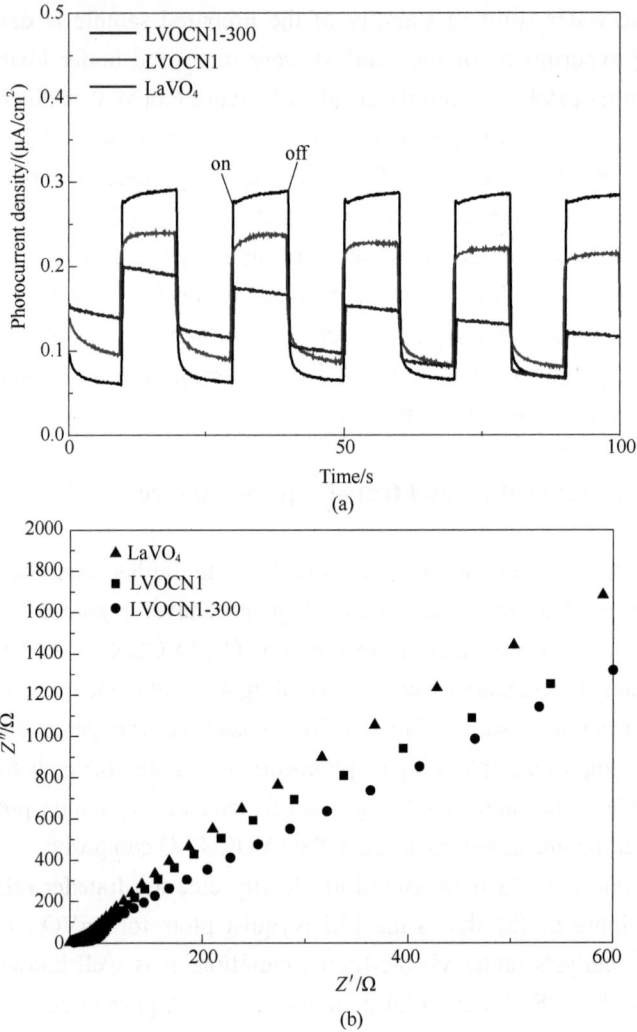

Figure 6.7　(a) Transient photocurrent density of $LaVO_4$, LVOCN1 and LVOCN1-O electrodes at 0.3 V versus Ag/AgCl, (b) electrochemical impedance spectra of the as-prepared samples at -0.4 V versus Ag/AgCl

6.4.3　Mechanism of water splitting

It has been widely concerned that the coupling two semiconductors method can form a heterojunction structure to improve the activity of photocatalysts. Based on the above characterization analysis and conclusions, we have an inference about the possible mechanism of the heterostructure LaVO$_4$/g-C$_3$N$_4$ composite to enhance the photocatalytic activity. It is well known that the effective separation of photoinduced electrons and holes can improve photocatalytic performance. As illustrated in Figure 6.8, the bandgap of LaVO$_4$/g-C$_3$N$_4$ composite is smaller than that of pure LaVO$_4$. Figure 6.9 shows that the electron-hole pairs in both lanthanum vanadate and carbon nitride will be excited under visible light irradiation, and the generated electron-hole pairs can migrate through the heterojunction interface of the two semiconductor materials at the appropriate position of VB and CB. Besides, the band gap energies of CB and VB of g-C$_3$N$_4$ are −1.22 and +1.56 eV approximately, while the CB and VB potentials of LaVO$_4$ are around −0.06 and +2.01 eV, respectively. It is obvious that the CB of g-C$_3$N$_4$ is more negative than that of LaVO$_4$. Therefore, owing to a sufficiently potential difference, photogenerated electrons could be readily transferred from g-C$_3$N$_4$ to CB of LaVO$_4$. Meanwhile, the great difference in VB potential of LaVO$_4$ and g-C$_3$N$_4$ leads to the photoinduced holes on the LaVO$_4$ surface moving towards g-C$_3$N$_4$, which reduces the possibility of electron-hole recombination, and generates more electrons on the LaVO$_4$ and g-C$_3$N$_4$ surfaces to further enhance the activity of photocatalyst. The electron

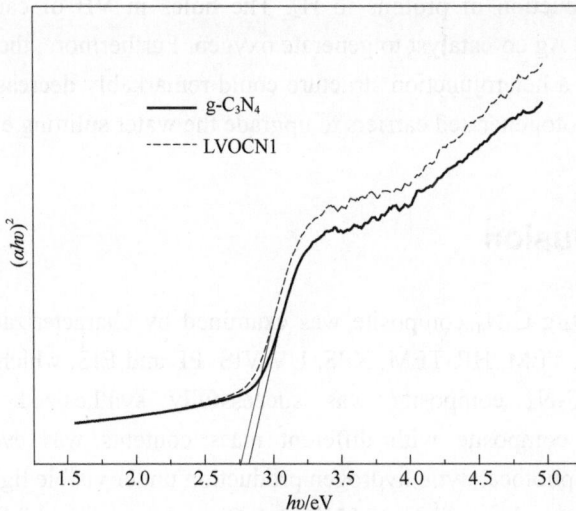

Figure 6.8   Plots of $(\alpha h\upsilon)^2$ versus photon energy $(h\upsilon)$ for the band gap energies of g-C$_3$N$_4$ and LVOCN1

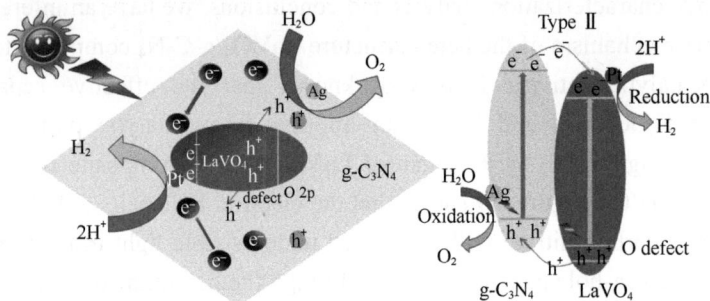

Figure 6.9   Schematic illustrations of the proposed photocatalytic mechanism over the prepared photocatalyst under visible light irradiation

transfer process will be expedited and the interface energy will be dropped by the heterojunction structure of the $LaVO_4$/g-C_3N_4 composite. The photocatalytic water splitting mechanism of the $LaVO_4$/g-C_3N_4 composite is speculated to be type II mechanism. From the mechanism of Figure 6.8, the electrons on the CB of g-C_3N_4 will be transferred to the CB of $LaVO_4$, and the holes on the VB of $LaVO_4$ via the O defect will be transferred to the VB of g-C_3N_4. The electrons accumulated on the CB of lanthanum vanadate that are further transferred to the Pt nanoparticles could catalyze the reduction of protons to H_2. The holes in VB of carbon nitride are interaction with Ag co-catalyst to generate oxygen. Furthermore, the $LaVO_4$/g-C_3N_4 composite with a heterojunction structure could remarkably decrease the migration resistance of photogenerated carriers to upgrade the water splitting efficiency.

6.5   Conclusion

The $LaVO_4$/g-C_3N_4 composite was examined by characterization technology including XRD, TEM, HR-TEM, XPS, UV-VIS, PL and EIS, which confirmed that the $LaVO_4$/g-C_3N_4 composite was successfully synthesized. The prepared $LaVO_4$/g-C_3N_4 composite with different mass contents was evaluated by the performance of photocatalytic hydrogen production under visible light irradiation. It is obvious that oxygen-defect LVOCN1-O composite exhibits the highest photocatalytic activity for H_2 and O_2 evolution rate of 0.89 mmol/(h·g) and 0.23 mmol/(h·g) under visible light illumination, about 27% and 35% higher than those of LVOCN1, respectively. The photocatalytic water splitting activity of LVOCN1 still displays satisfactory stability after four recycling runs. Schematic illustrations of mechanism indicate that Type II mechanism may be a possible mechanism for photocatalytic water splitting reactions. The photogenerated electrons can be easily transferred from g-C$_3$N$_4$ to the CB of LaVO$_4$, which indicates that the photoinduced electron-hole pairs are easily separated. It can be indicated that photocatalytic water splitting activity of LaVO$_4$/g-C$_3$N$_4$ composite can be upgraded with oxygen defects.

Reference

[1] Song M., Wu Y., Zheng G., et al. Junction of porous g-C$_3$N$_4$ with BiVO$_4$ using Au as electron shuttle for cocatalyst-free robust photocatalytic hydrogen evolution[J]. Applied Surface Science, 2019, 498.

[2] Luo Y., Deng B., Pu Y., et al. Interfacial coupling effects in g-C$_3$N$_4$/SrTiO$_3$ nanocomposites with enhanced H$_2$ evolution under visible light irradiation[J]. Applied Catalysis B: Environmental, 2019, 247: 1-9.

[3] Liu X., Liu L., Yao Z., et al. Enhanced visible-light-driven photocatalytic hydrogen evolution and NO photo-oxidation capacity of ZnO/g-C$_3$N$_4$ with N dopant[J]. Colloids and Surfaces A: Physicochemical and Engineering Aspects, 2020, 599: 124869.

[4] Xu Y., Liu J., Xie M., et al. Construction of novel CNT/LaVO$_4$ nanostructures for efficient antibiotic photodegradation. Chemical Engineering Journal, 2019, 357:487-497.

[5] Wang S., Teng Z., Xu Y., et al. Defect as the essential factor in engineering carbon-nitride-based visible-light-driven Z-scheme photocatalyst[J]. Applied Catalysis B: Environmental, 2020, 260: 118145.

[6] Wang J., Xia Y., Zhao H.,et al. Oxygen defects-mediated Z-scheme charge separation in g-C$_3$N$_4$/ZnO photocatalysts for enhanced visible-light degradation of 4-chlorophenol and hydrogen evolution[J]. Applied Catalysis B: Environmental, 2017, 206: 406-416.

[7] Yuan Y. J., Shen Z. K., Wang P., et al. Metal-free broad-spectrum PTCDA/g-C$_3$N$_4$ Z-scheme photocatalysts for enhanced photocatalytic water oxidation[J]. Applied Catalysis B: Environmental, 2020, 260: 118179.

[8] Liu X., Zhang Q., Liang L., et al. In-situ growing of CoO nanoparticles on g-C$_3$N$_4$ composites with highly improved photocatalytic activity for hydrogen evolution[J]. Royal Society Open Science, 2019, 6 (7): 190433.

[9] Gao D., Xu Q., Zhang J., et al. Defect-related ferromagnetism in ultrathin metal-free g-C$_3$N$_4$ nanosheets[J]. Nanoscale, 2014, 6 (5): 2577-2581.

[10] Liu X., He L., Chen X., et al. Facile synthesis of CeO$_2$/g-C$_3$N$_4$ nanocomposites with significantly improved visible-light photocatalytic activity for hydrogen evolution[J].

International Journal of Hydrogen Energy, 2019, 44 (31): 16154-16163.

[11] Wang S., Li C., Wang T., et al. Controllable synthesis of nanotube-type graphitic C_3N_4 and their visible-light photocatalytic and fluorescent properties[J].Journal of Materials Chemistry A, 2014, 2 (9): 2885-2890.

[12] Veldurthi N. K., Eswar N. K., Singh S. A., et al. Cocatalyst free Z-schematic enhanced H_2 evolution over $LaVO_4/BiVO_4$ composite photocatalyst using Ag as an electron mediator[J]. Applied Catalysis B: Environmental, 2018, 220: 512-523.

[13] He Y., Cai J., Zhang L., et al. Comparing two new composite photocatalysts, t-$LaVO_4$/g-C_3N_4 and m-$LaVO_4$/g-C_3N_4, for their structures and performances[J].Industrial & Engineering Chemistry Research, 2014, 53 (14): 5905-5915.

Chapter 7

Co-C$_3$N$_4$/BiPO$_4$ composite

7.1 Background

The photoactivity of Z-scheme g-C$_3$N$_4$ based composites for water splitting is developed by the fast charge transportation, effective photogenerated electron-hole pairs separation as well as high solar-light absorbance[1]. The Z-scheme g-C$_3$N$_4$ based photocatalysts, such as MnO$_2$/g-C$_3$N$_4$[2], Cu$_2$O/g-C$_3$N$_4$[3], NiCo$_2$O$_4$/g-C$_3$N$_4$[4], CeO$_2$/g-C$_3$N$_4$[5-7], N-ZnO/g-C$_3$N$_4$[8], Cr$_2$O$_3$/g-C$_3$N$_4$[9], Mn-Fe$_2$O$_3$/g-C$_3$N$_4$[10], are established with noble metals to improve photocatalytic water-splitting activity. As the scarce noble metals with high cost are not the ideal materials for water-splitting, the non-noble materials with solar light sensitive and earth-abundant need to be explored.

The traditional transition metals are abundant and eco-friendly, which have already attracted more attention for photocatalytic water-splitting. Fu reported that the prepared NiO/g-C$_3$N$_4$ composite had high photocatalytic performance for water splitting[11]. It has been reported that Co based material had the photocatalytic activity for overall water splitting[12]. Co combined with g-C$_3$N$_4$ can be used as the effective photocatalyst to improve the photocatalytic water splitting activity[13]. However, the Co/g-C$_3$N$_4$ composite can be easily poisoned by generated H$_2$O$_2$ during the photocatalytic overall water splitting reaction to cause rapid inactivation[14]. To overcome this challenge, some efforts have been devoted to the efficient photocatalytic overall water splitting system without noble metals and sacrificial reagents. Co combined with the reduced graphene oxide (RGO) as the heat conductor shows excellent photocatalytic stability of water splitting[15]. For C$_3$N$_4$ material, it is reported that the lattice-matched material is combined with C$_3$N$_4$ to establish core/shell structures of C$_3$N$_4$ based composites, which can obtain the stable photocatalytic activity[16].

Bismuth phosphate (BiPO$_4$), has been reported as an oxoacid salt material with substantial chemical stability to be active for photocatalytic performance[17,18]. The enhanced photocatalytic performance is achieved by the match of lattice and energy level of C$_3$N$_4$/BiPO$_4$ composite with the efficient separation of photogenerated electron-hole pairs[16]. Besides, the band gap energy of the conduction band of BiPO$_4$ is about -0.7 V vs SCE, and it is considerably more negative than the reduction potential of water to be a potential photocatalyst for H$_2$ evolution[19].

In this book, core/shell structures of Co-C$_3$N$_4$/BiPO$_4$ photocatalyst are prepared to improve photocatalytic water-splitting performance. The physical, chemical and photoelectric properties of Co-C$_3$N$_4$/BiPO$_4$ composite with different Co-C$_3$N$_4$ loading are studied by a series of characterizations. The enhanced mechanism of photocatalytic H$_2$ evolution activity from water-splitting of the as-prepared Co-C$_3$N$_4$/BiPO$_4$ composite is also discussed.

7.2 Preparation of Co-C$_3$N$_4$/BiPO$_4$ composite

Urea, cobalt nitrate, Bi(NO$_3$)$_3 \cdot$6H$_2$O, Na$_3$PO$_4 \cdot$12H$_2$O, nitric acid and ethanol were purchased from Aladdin Industrial Corporation with analytical grade.

The g-C$_3$N$_4$ was synthesized by heating urea for 3 h in the covered crucible at 500 °C in a muffle furnace. The pure BiPO$_4$ was prepared via the hydrothermal method. Three mmol of Na$_3$PO$_4 \cdot$12H$_2$O, Bi(NO$_3$)$_3 \cdot$6H$_2$O and 30 mL of distilled water were placed into a beaker. Then, the concentrated HNO$_3$ was added to the solution to adjust PH to 1 and magnetically stirred for 1 h. The solution was subsequently put into the 100 mL Teflon-lined stainless steel autoclave and kept the hydrothermal temperature at 160 °C for 12 h. After the hydrothermal process, the sediments were washed with distilled water for several times and dried at 80 °C to obtain BiPO$_4$.

Co-g-C$_3$N$_4$ composite was synthesized by the facile thermal method. Typically, 2 mL Co(NO$_3$)$_2$ (2 mg/mL) aqueous solution was mixed with 200 mg C$_3$N$_4$ and dispersed in the pure water by sonication. The obtained mixture was kept stirring in the water bath at 70 °C for one night and was rotavaporated to dryness. The remainder was calcined in the tube furnace at 400 °C under a N$_2$ atmosphere for 2 h to form Co-C$_3$N$_4$ samples.

The Co-C$_3$N$_4$/BiPO$_4$ photocatalyst was prepared by the self-assembly formation method. Co-C$_3$N$_4$ composite was dispersed with ultrasonic waves in an ethanol solvent and the as-synthesized BiPO$_4$ powders were put into this solvent. The mixture was stirred for one hour and then placed into an air dry oven at 75 °C for 2 h. The Co-C$_3$N$_4$ composite could be spontaneously coated on BiPO$_4$ during the evaporation process. The Co-C$_3$N$_4$/BiPO$_4$ samples were prepared by this method and named as x-Co-CN/BP, where x represents the mass ratio of Co-C$_3$N$_4$ to BiPO$_4$ (1%, 3%, 5%, 7% and 10%).

7.3   Characterization of Co-C$_3$N$_4$/BiPO$_4$ composite

7.3.1   XRD of Co-C$_3$N$_4$/BiPO$_4$ composite

The crystalline structures of Co-CN/BiPO$_4$ composite, Co-C$_3$N$_4$ and pure BiPO$_4$ were investigated by XRD, and the results are shown in Figure 7.1 . The reflection of the Co-C$_3$N$_4$ sample shows two typical peaks of g-C$_3$N$_4$ at 13.3° and 27.4°[20]. There are no reflections of Co crystalline structure in comparison with fresh g-C$_3$N$_4$ showns in Figure 7.2, indicating that the Co nanoparticles are

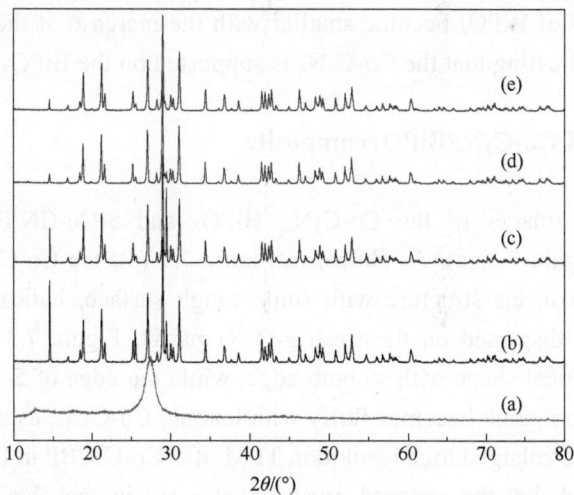

Figure 7.1   XRD patterns of Co-C$_3$N$_4$, BiPO$_4$, 3-Co-CN/BP, 5-Co-CN/BP and 10-Co-CN/BP composite

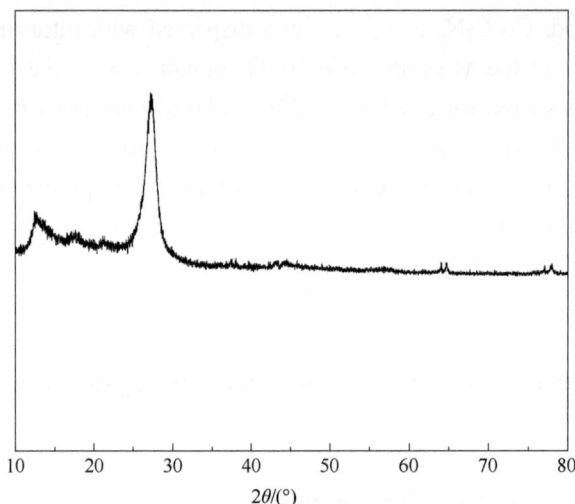

Figure 7.2　XRD pattern of g-C$_3$N$_4$

homogeneously growing in the C$_3$N$_4$ sheets. The crystalline structures of the prepared BiPO$_4$ based materials are well and the diffraction peaks can be perfectly matched with the BiPO$_4$ with hexagonal structure (JCPDS 15-0766), in agreement with XRD patterns of BiPO$_4$ reported by Lin et al.[21]. The diffraction peaks at 14.6° and 27.1° of BiPO$_4$ become smaller with the increase of the mass contents of Co-C$_3$N$_4$, indicating that the Co-C$_3$N$_4$ is supported on the BiPO$_4$ layers.

7.3.2　TEM of Co-C$_3$N$_4$/BiPO$_4$ composite

The TEM images of the Co-C$_3$N$_4$, BiPO$_4$ and 5-Co-CN/BP samples are displayed in Figure 7.3 (a). As shown in Figure 7.3 (a), the Co-C$_3$N$_4$ sample has multi-layer nanosheets structure with some rough surface, indicating that Co is homogeneously dispersed on the fresh g-C$_3$N$_4$ matrix. Figure 7.3 (b) shows that BiPO$_4$ is cylindrical shape with smooth edge, while the edge of 5-Co-CN/BP with self-assembly formation becomes fluffy with loading Co-C$_3$N$_4$, as shown in Figure 7.3 (c). From the enlarged high resolution TEM of 5-Co-CN/BP in Figure 7.3 (d), it can be observed that the exposed crystal planes are in the obtained Co-CN/BP composite. The lattice fringe with the spacing of 0.311 nm is corresponds to the

(002) planes of C_3N_4. The lattice fringes with the spacing of 0.313 nm and 0.42 nm are attributed to the (200) and (011) planes of C_3N_4. The lattice-matched heterojunction structure is formed between the C_3N_4 and $BiPO_4$, as shown in Figure 7.4 (a). To further display the composition of the Co-CN/BP composite, the EDS mapping images of 5-Co-CN/BP composite are shown in Figure 7.4 (b). As can be seen from the images, the Bi and P elements are surrounded by C and N elements, and cobalt is uniformly distributed over the composite. Based on these results, it is confirmed that the Co-CN/BP composite with lattice-matched heterojunction structure is successfully prepared.

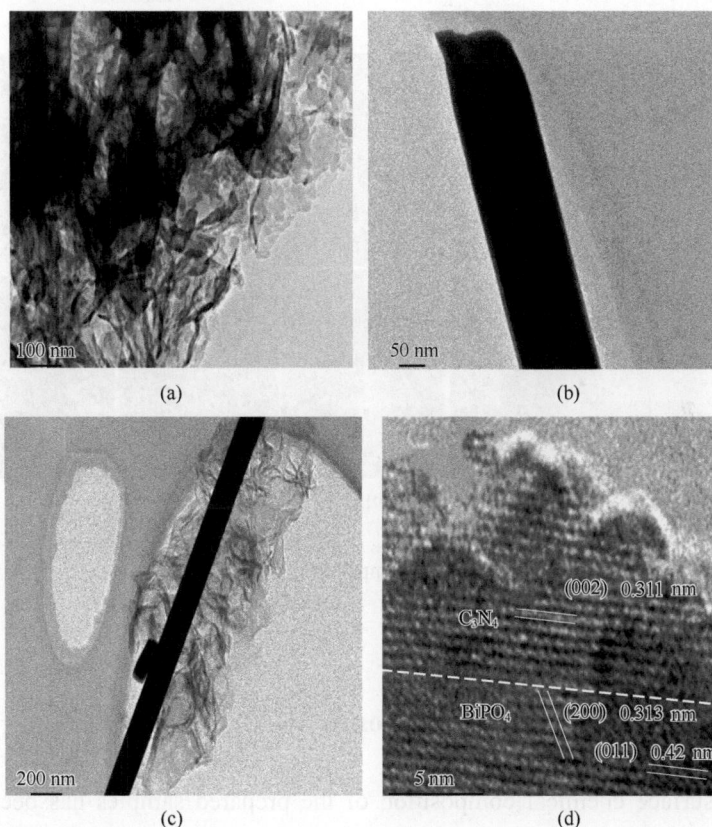

Figure 7.3   TEM images of (a) Co/C_3N_4, (b) $BiPO_4$, (c) and (d) 5-Co-CN/BP composite

(a)

(b)

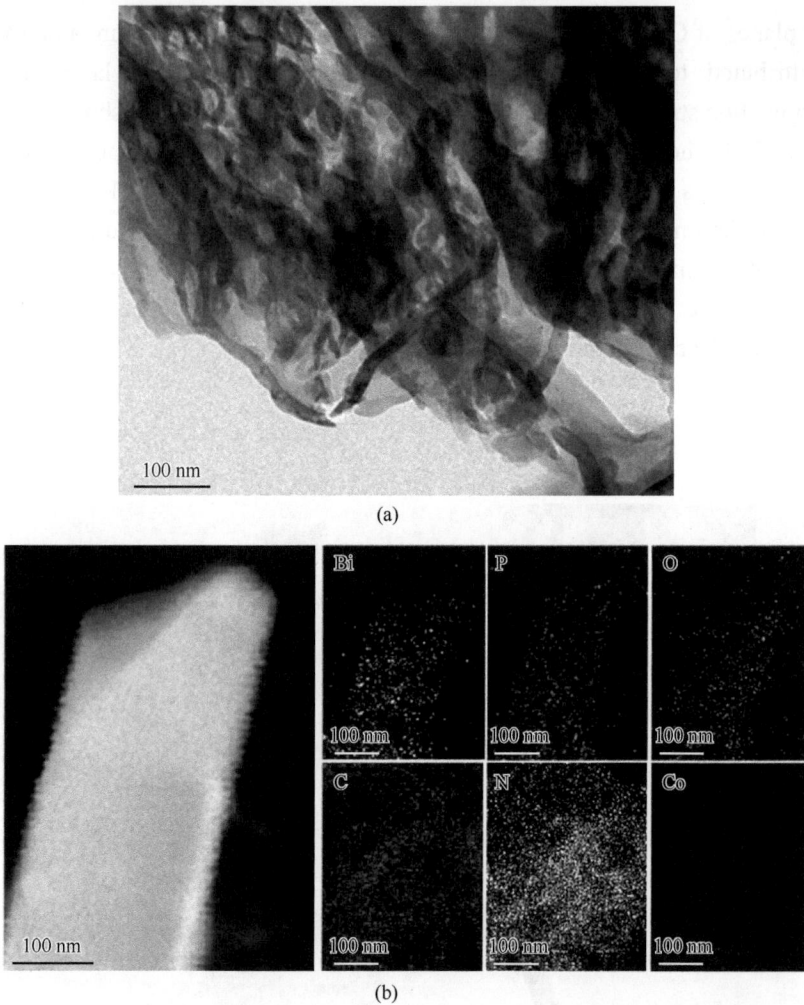

Figure 7.4 (a) TEM image of fresh g-C$_3$N$_4$, (b) The EDS mapping images of 5-Co-CN/BP composite

7.3.3 XPS of Co-C$_3$N$_4$/BiPO$_4$ composite

The surface chemical composition of the prepared samples has been deeply characterized by XPS, and the XPS with the results displayed in Figure 7.5 (a)-(f). Figure 7.5(a) shows the full survey spectrum of 5-Co-CN/BP and 10-Co-CN/BP composite compared to that of pure Co-C$_3$N$_4$ and BiPO$_4$. The atomic relative contents (%) of prepared samples are presented in Table 7.2. The sharp peaks at binding energy values of around 159 eV, 131 eV, 399 eV, 530 eV, 285 eV and 782 eV are attributed to the signals of Bi 4f, P 2p, N 1s, O 1s, C 1s and Co 2p, respectively in the Co-CN/BP composite. From BiPO₄ to 10-Co-CN/BP, the contents of Bi, P and O decrease, while the content of N increases after Co-C₃N₄ coating. The N 1s spectra shows in Figure 7.5(b) can be deconvoluted into four peaks at about 398.5 eV, 399 eV, 400.8 eV, and 404.7 eV, respectively. The binding energy at 398.5 eV can be ascribed to the signals of the sp^2-hybridized nitrogen atoms in C=N—C groups[22]. The binding energy at about 399 eV is attributed to the tertiary nitrogen N—C₃ groups or H—N—C₂[23]. The binding energy at 400.8 eV is corresponding to the amino function groups[22], while the peak observed at 404.7 eV results from charging effects or positive charge localization in heterocycles[24]. The binding energy and relative contents of N 1s for the Co-C₃N₄, 5-Co-CN/BP and 10-Co-CN/BP composite are shown in Table 7.1. It can be seen that the banding energy of N 1s spectra of the 5-Co-CN/BP and 10-Co-CN/BP composite shows a negative shift compared with that of Co-C₃N₄. On the contrary, the banding energy of Bi 4f in Figure 7.5(c) and P 2p in Figure 7.5(d) of Co-CN/BP composite exhibits the positive shift in comparison with that of pure BiPO₄. These results confirm that the Co-C₃N₄ nanocomposite is chemical coated on the BiPO₄ nanorod rather than physical adsorption. Figure 7.5(f) is presented the high-resolution XPS spectra of

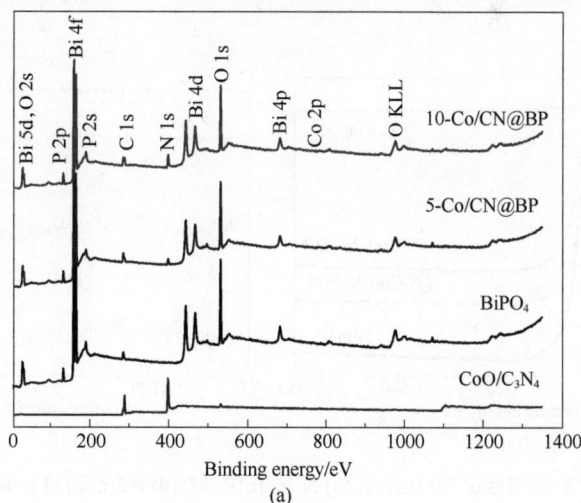

Figure 7.5

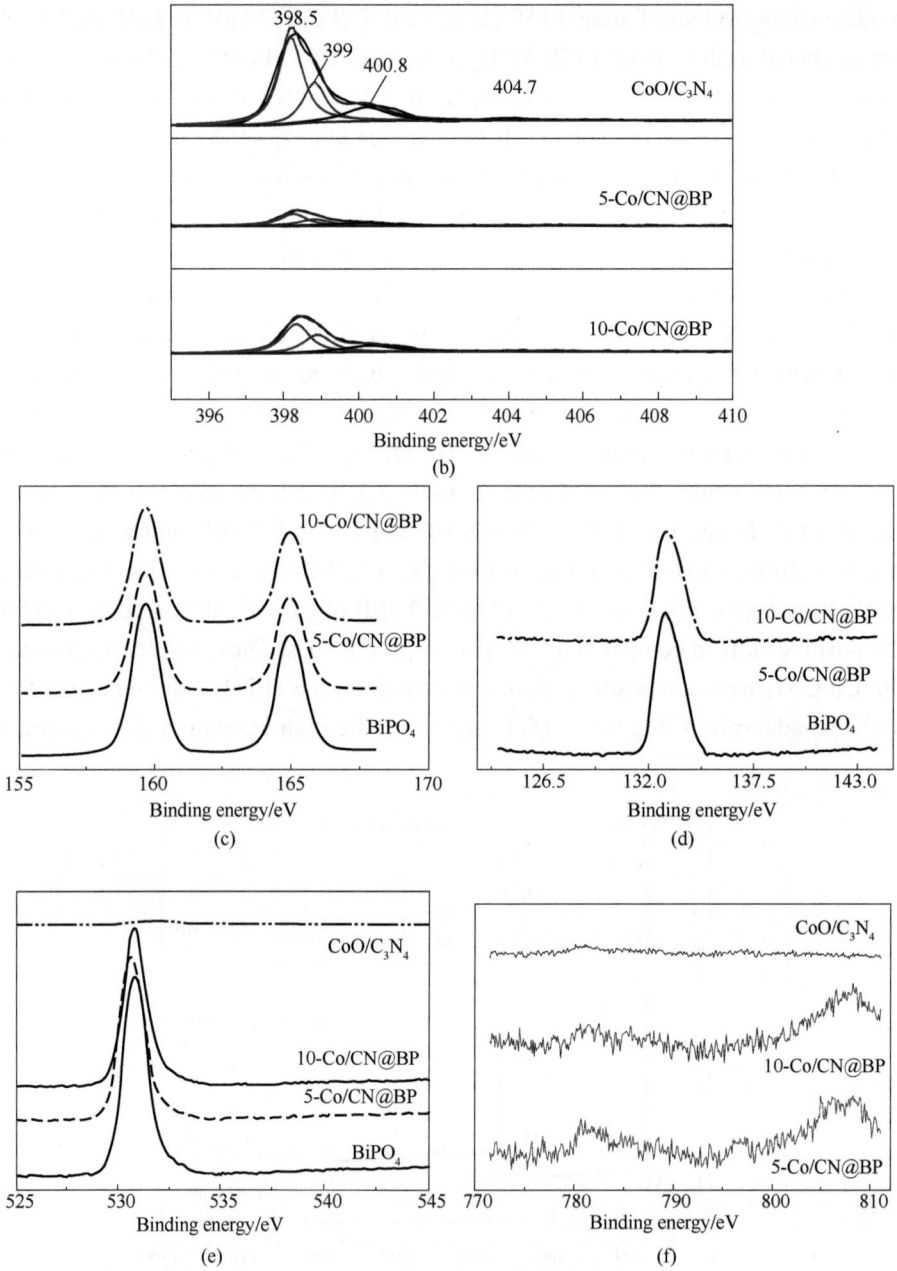

Figure 7.5   XPS profiles of (a) survey, (b) N 1s, (c)Bi 4f, (d) P 2p, (e) O 1s and (f) Co 2p of the prepared samples

Co 2p spectra. The Co 2p spectra with the weak and diffused peaks at 780.3 eV and 796.2 eV of Co-CN/BP composite confirm the existence of Co, which are ascribed to the major binding energies of Co^{2+} in $Co^{[25]}$. It is observed that the intensity of the Co 2p peaks in the 5-Co-CN/BP composite is strongest and the banding energy of O 1s spectra in Figure 7.5(e) of the 5-Co-CN/BP composite shows the most positive shift in the prepared materials, indicating the strong interaction between Co and O in the 5-Co-CN/BP composite.

Table 7.1   BE and RC of N 1s for prepared samples

Sample	Co/C₃N₄		10-Co-CN/BP		5-Co-CN/BP	
	BE/eV	RC/%	BE/eV	RC/%	BE/eV	RC/%
N=C—N	398.21	58.51	398.28	46.77	398.35	40.79
N—C₃	398.84	25.32	398.90	30.42	398.92	25.97
Amino function	400.22	16.16	400.32	20.06	400.35	21.90
Charging effects	404.70	0.01	405.08	2.75	405.40	11.34

Table 7.2.   Atomic relative content (%) of prepared samples from XPS characterization

Sample	C₃N₄	Co-C₃N₄	5-Co-CN/BP	10-Co-CN/BP	BiPO₄
N	52.66	50.43	7.65	15.99	—
C	44.28	46.27	24.59	24.91	—
O	3.05	2.31	47.03	40.35	68.36
Co	—	0.99	0.15	0.35	—
P	—	—	12.13	10.70	18.56
Bi	—	—	8.45	7.70	13.08

7.3.4   UV-vis absorption and PL spectra of Co-C₃N₄/BiPO₄ composite

UV-vis absorption spectra and photoluminescence (PL) are carried out to determine the optical and photoelectric properties of BiPO₄ and Co-CN/BP composite. Figure 7.6 (a) shows the UV-vis diffuse reflectance spectra of fresh BiPO₄, 3-Co-CN/BP, 5-Co-CN/BP and 10-Co-CN/BP composite. It can be observed that 5-Co-CN/BP has the best visible-light and ultraviolet absorbance, which could utilize more solar light to enhance the photocatalytic activity for hydrogen evolution. The band gap energies of BiPO₄, 5-Co-CN/BP and Co-C₃N₄ were calculated from the formula $(\alpha h\upsilon)^n = k(h\upsilon - E_g)$, where α, k, $h\upsilon$ and E_g were absorption coefficient, a constant related to the effective masses associated with the conduct band and valence band, the absorption energy and band gap energy, respectively, and the value of parameter n was 2 for direct-allowed transition. The E_g values of $BiPO_4$, 5-Co-CN/BP and Co-C_3N_4 composite in Figure 7.6 (b) were 4.46 eV, 4.29 eV and 2.94 eV, respectively. The PL spectra of the prepared samples with excitation wavelength of 370 nm are displayed in Figure 7.6 (c). It is seen that the emission profile decreases from $BiPO_4$ to 10-Co-CN/BP composite, which indicates that the separation rate of the photogenerated charge carrier becomes faster by the Co-C_3N_4 loading. It is well known that the efficient separation of photo-induced electron-hole pairs is more beneficial to enhance the photocatalytic activity[26].

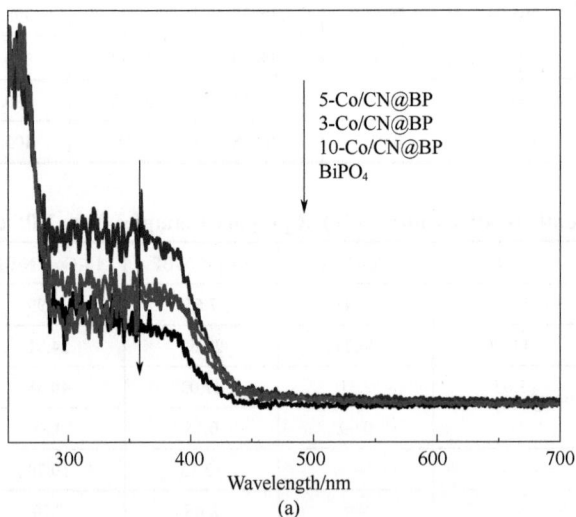

5-Co/CN@BP
3-Co/CN@BP
10-Co/CN@BP
$BiPO_4$

Wavelength/nm

(a)

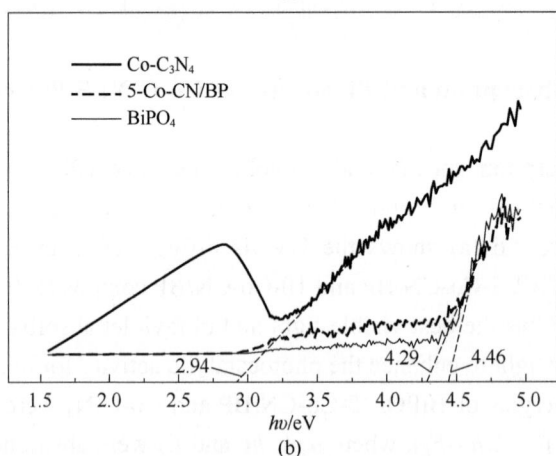

Co-C_3N_4
5-Co-CN/BP
$BiPO_4$

2.94   4.29   4.46

$h\upsilon$/eV

(b)

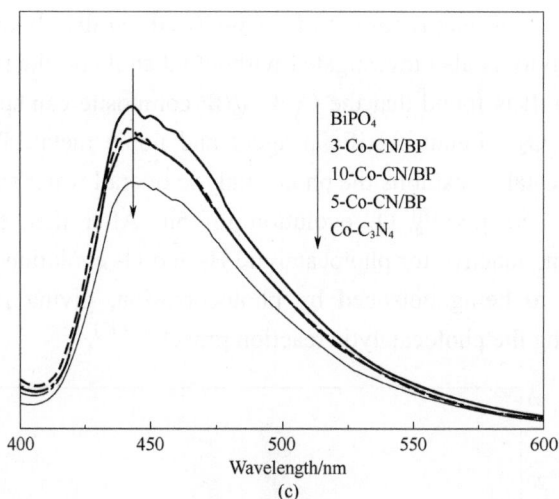

Figure 7.6   (a) UV-vis absorption spectra of $BiPO_4$, 3-Co-CN/BP, 5-Co-CN/BP and 10-Co-CN/BP composite, (b) Plots of $(\alpha h\upsilon)^2$ versus photon energy (hυ) for the band gap energies of $BiPO_4$, 5-Co-CN/BP and Co-C_3N_4, (c) PL spectra (λ_{ex} = 370 nm) for the prepared $BiPO_4$, 3-Co-CN/BP, 5-Co-CN/BP, 10-Co-CN/BP and Co-C_3N_4 composite

7.4   Photocatalytic water splitting testing Co-C₃N₄/BiPO₄ composite

7.4.1   Water splitting efficiency

The photocatalytic H_2 evolution activity of the prepared 5-Co-CN/BP composite is measured without noble metal and the results are displayed in Figure 7.7. In Figure 7.7 (a), the photocatalytic H_2 evolution rate for $BiPO_4$, 1-Co-CN/BP, 3-Co-CN/BP 5-Co-CN/BP, 7-Co-CN/BP, 10-Co-CN/BP, Co-C_3N_4 and C_3N_4 are calculated to be 8.3 μmol/(h·g), 11.0 μmol/(h·g), 15.8 μmol/(h·g), 23.9 μmol/(h·g), 22.8 μmol/(h·g), 18.5 μmol/(h·g), 14.3 μmol/(h·g) and 1.4 μmol/(h·g), respectively. The 5-Co-CN/BP exhibits a superior photocatalytic hydrogen evolution rate, about two times higher than that of fresh $BiPO_4$. Figure 7.7 (b) displays the photocatalytic stability of H_2 evolution for the 5-Co-CN/BP composite is carried out by three cyclic experiments. The 5-Co-CN/BP composite exhibits well stability for photocatalytic H_2 evolution from water-splitting with three recycling

111

runs. The influence of the proportion of 5-Co-CN/BP on the photocatalytic H_2 and O_2 evolution activity is also investigated without ethanol and the results are shown in Figure 7.7 (c). It is found that the Co-CN/BP composite can split pure water to produce H_2 and O_2 without sacrificial agent and noble metal. The 5-Co-CN/BP composite photocatalyst exhibits the photocatalytic overall water splitting with 3.92 µmol/g H_2 and 1.87 µmol/g O_2 evolution for 5h. After that, the 5-Co-CN/BP composite exhibits inactive for photocatalytic H_2 and O_2 evolution from pure water splitting, owing to being poisoned by photocorrosion, giving rise to the rapid inactivation during the photocatalytic reaction process[14,27].

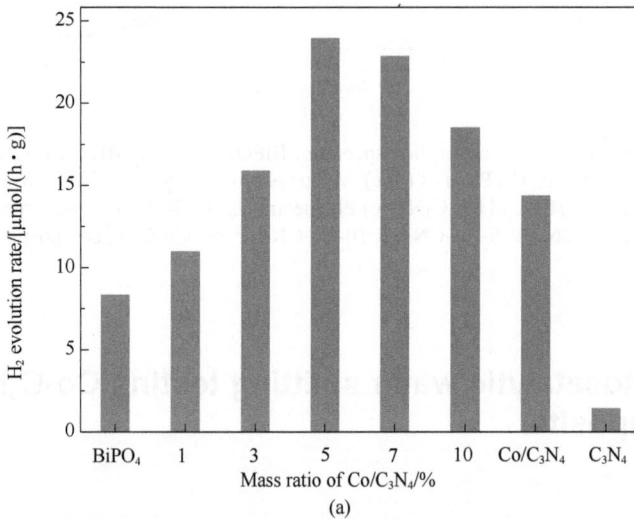

(a)

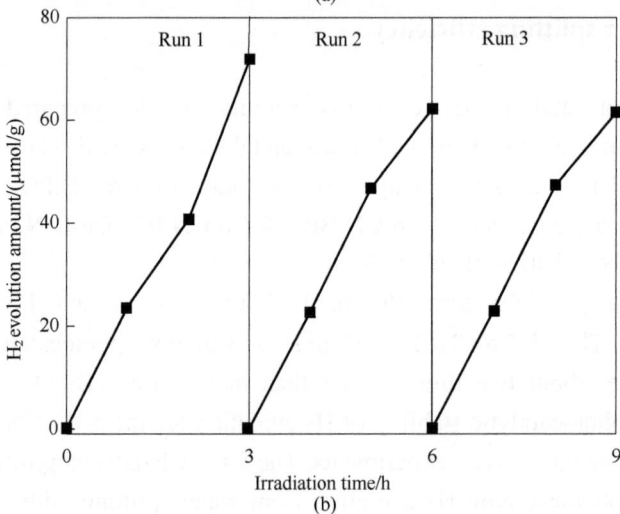

(b)

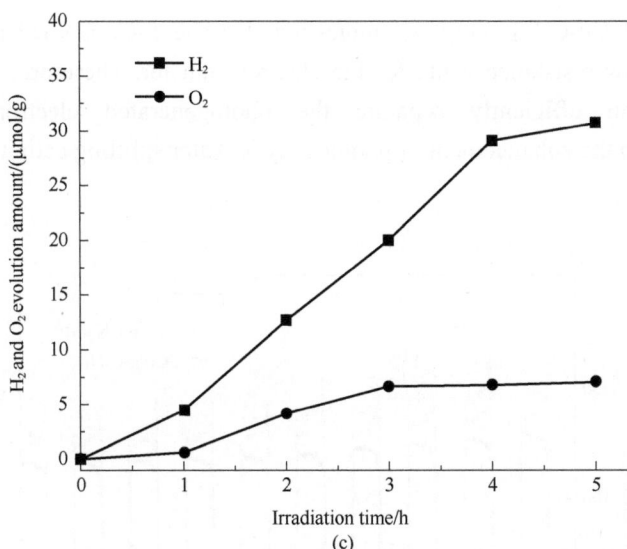

Figure 7.7   (a) the photocatalytic H_2 evolution rate of $BiPO_4$, Co-CN/BP, Co-C₃N₄ and C₃N₄, (b) recyclability of 5-Co-CN/BP photocatalyst for the photocatalytic H_2 evolution, (c) the amount of photocatalytic H_2 and O_2 evolution of 5-Co-CN/BP without sacrificial agent

7.4.2   Charge separation and transfer performance

The photoelectrochemical measurements including electrochemical impedance spectroscopy (EIS) and photocurrents are carried out to explore the photogenerated charge separation and transfer properties of $BiPO_4$, Co-CN/BP and Co-C₃N₄ samples. Figure 7.8 (a) displays the transient photocurrent responses for $BiPO_4$, 5-Co-CN/BP and Co-C₃N₄ samples at the interval light-on and light-off. It can be seen that 5-Co-CN/BP composite presents the highest transient photocurrent response, suggesting that the faster interfacial electron transfer and better separation efficiency of electron-hole pairs than that of pure $BiPO_4$ and Co-C₃N₄ samples. The electrochemical impedance spectroscopy (EIS) measurement is also the powerful method to evaluate the photogenerated electron transfer in the samples. Figure 7.8 (b) shows the Nyquist impedance plots for $BiPO_4$, 5-Co-CN/BP and Co-C₃N₄ samples. It is well known that the smaller arc radius attributes to the faster interfacial electron transfer and better separation efficiency of electron-hole pairs. The Nyquist plot of 5-Co-CN/BP exhibits the obvious smaller arc diameter than that

113

of $BiPO_4$ and $Co-C_3N_4$ samples, suggesting that the 5-Co-CN/BP photocatalyst possesses a low resistance value for the electron transfer. Therefore, 5-Co-CN/BP composite can efficiently separate the photogenerated electronhole pairs, contributing to the enhancement of photocatalytic water splitting activity.

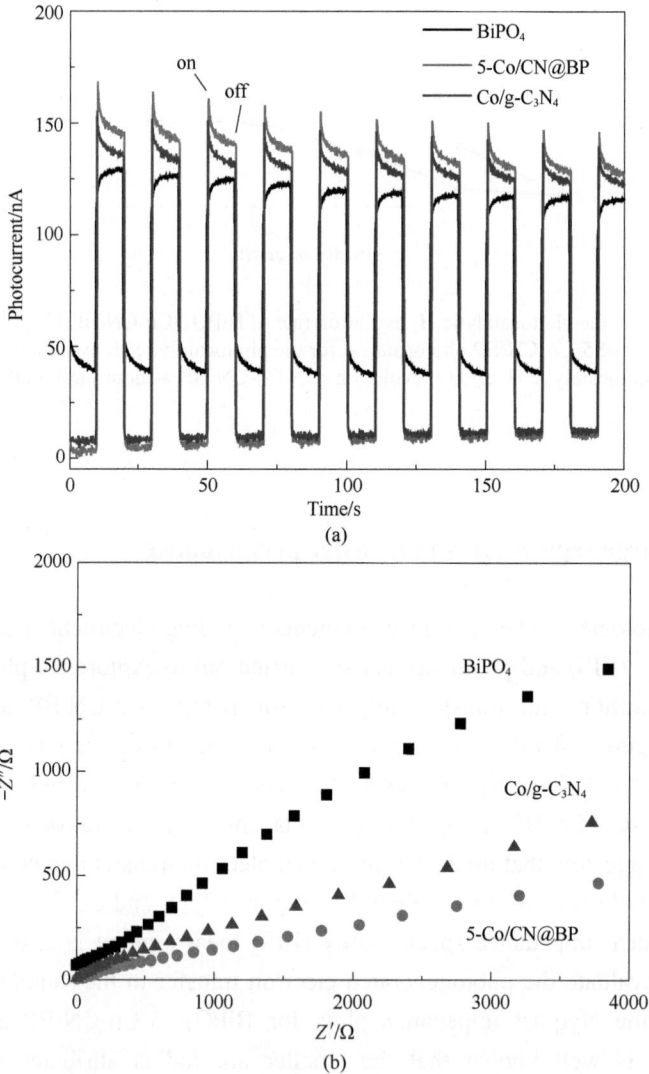

Figure 7.8 (a) Transient photocurrents, (b) electrochemical impedance spectra of $BiPO_4$, 5-Co-CN/BP and Co-g-C_3N_4 electrodes at 0.3 V and −0.4 V versus Ag/AgCl

7.4.3   Mechanism of hydrogen production

Based on the above description and discussion, the possible Z-scheme photocatalytic mechanism of overall water splitting for the Co-CN/BP photocatalyst is proposed, as shown in Figure 7.9 (b). First, the electron-hole pairs will be generated in the CB and VB of BiPO$_4$ and g-C$_3$N$_4$ by light irradiation. The photo-excited electron-hole pairs will migrate through the lattice-matched heterojunction interface of two semiconductor materials with suitable positions of VB and CB. The band gap energies of CB and VB of g-C$_3$N$_4$ are about −1.22 eV and +1.56 eV, while the CB and VB potentials of BiPO$_4$ are about −0.26 eV and +4.2 eV[21], respectively. The CB of BiPO$_4$ is lower than that of g-C$_3$N$_4$. The photo-excited electrons on the VB of BiPO$_4$ can easily migrate to the VB of g-C$_3$N$_4$ by the potential difference, thus recombination of electrons and holes could be effectively limited. The remaining electrons on the CB of g-C$_3$N$_4$ can further transfer to Co nanoparticles to catalyze the reduction of protons to generate hydrogen[28]. As the VB potential of BiPO$_4$ (+4.2 eV vs. NHE) is more positive than the standard redox potential E(O$_2$/H$_2$O) (+1.23 eV vs. NHE), the holes in the VB of BiPO$_4$ will react with H$_2$O to generate O$_2$[26]. As shown in the Figure 7.9 (a), Co

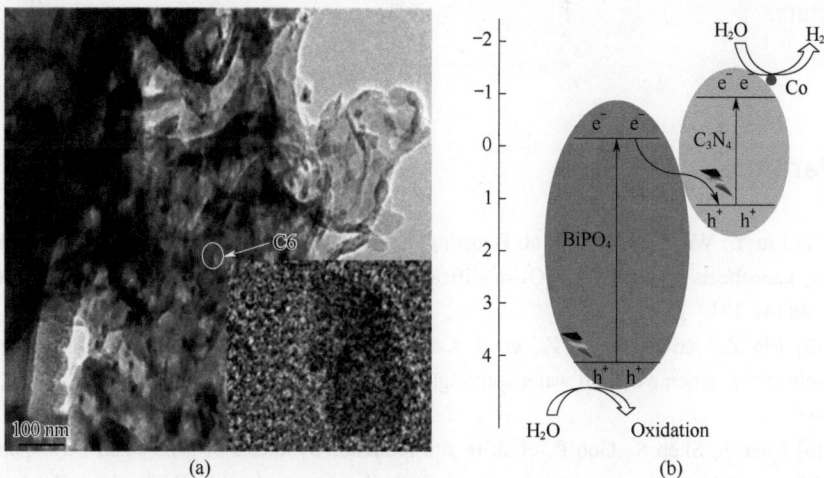

Figure 7.9   (a) HR-TEM image of Co-C$_3$N$_4$, (b) Schematic illustrations of the proposed photocatalytic water-splitting mechanism of the Z-scheme Co-CN/BP photocatalyst

nanoparticles existed on the g-C_3N_4 surface rather than dispersed on the $BiPO_4$, demonstrating that photoinduced electrons in the CB of g-C_3N_4 remain on the original site as predicted by the Z-scheme route instead of transferring to the CB of $BiPO_4$ through the conventional type-II mechanism. Besides, the Co-CN/BP composite with the lattice-matched structure could greatly reduce the migration resistance of photogenerated carriers. The electron transfer process and the separation of photoinduced electrons and holes will be enhanced to improve the photocatalytic water splitting activity from overall water splitting by the lattice-matched heterojunction structure of Co-CN/BP photocatalyst.

7.5 Conclusion

The Co-CN/BP photocatalyst is facilely synthesized to investigate the photocatalytic H_2 evolution activity from water splitting. The Co-CN/BP photocatalyst shows well photocatalytic H_2 evolution activity from overall water splitting without noble metal.The Co-CN/BP composite with 5wt% Co-C_3N_4 loading exhibits the best photocatalytic activity for H_2 evolution rate of 23.9 $\mu mol/(h \cdot g)$. The greatly enhanced photocatalytic activity for H_2 evolution of Co-CN/BP composite is mainly owing to the faster charge transfer and more effective separation of electron-hole pairs by the lattice-matched heterojunction structure.

Reference

[1] Liu Y., Wu X., Lv H., et al. Boosting the photocatalytic hydrogen evolution activity of g-C_3N_4 nanosheets by $Cu_2(OH)_2CO_3$-modification and dye-sensitization[J]. Dalton Transactions, 2019, 48 (4): 1217-1225.

[2] Mo Z., Xu H., Chen Z., et al. Construction of MnO_2/Monolayer g-C_3N_4 with Mn vacancies for Z-scheme overall water splitting[J]. Applied Catalysis B: Environmental, 2019, 241: 452-460.

[3] Chen J., Shen S., Guo P., et al. In-situ reduction synthesis of nano-sized Cu_2O particles modifying g-C_3N_4 for enhanced photocatalytic hydrogen production[J]. Applied Catalysis B: Environmental, 2014, 152-153: 335-341.

[4] Chang W., Xue W., Liu E., et al. Highly efficient H_2 production over $NiCo_2O_4$ decorated

g-C$_3$N$_4$ by photocatalytic water reduction[J]. Chemical Engineering Journal, 2019, 362: 392-401.

[5] Zou W., Deng B., Hu X., et al. Crystal-plane-dependent metal oxide-support interaction in CeO$_2$/g-C$_3$N$_4$ for photocatalytic hydrogen evolution[J]. Applied Catalysis B: Environmental, 2018, 238:111-118.

[6] Liu X., He L., Chen X., et al. Facile synthesis of CeO$_2$/g-C$_3$N$_4$ nanocomposites with significantly improved visible-light photocatalytic activity for hydrogen evolution[J]. International Journal of Hydrogen Energy, 2019, 44 (31): 16154-16163.

[7] Zou W., Shao Y., Pu Y., et al. Enhanced visible light photocatalytic hydrogen evolution via cubic CeO$_2$ hybridized g-C$_3$N$_4$ composite[J]. Applied Catalysis B: Environmental, 2017, 218: 51-59.

[8] Liu Y., Liu H., Zhou H., et al. A Z-scheme mechanism of N-ZnO/g-C$_3$N$_4$ for enhanced H$_2$ evolution and photocatalytic degradation[J]. Applied Surface Science, 2019, 466: 133-140.

[9] Shi J., Cheng C., Hu Y., et al. One-pot preparation of porous Cr$_2$O$_3$/g-C$_3$N$_4$ composites towards enhanced photocatalytic H$_2$ evolution under visible-light irradiation[J]. International Journal of Hydrogen Energy, 2017, 42 (7): 4651-4659.

[10] Wang N., Han B., Wen J., et al. Synthesis of novel Mn-doped Fe$_2$O$_3$ nanocube supported g-C$_3$N$_4$ photocatalyst for overall visible-light driven water splitting[J]. Colloids and Surfaces A: Physicochemical and Engineering Aspects, 2019, 567: 313-318.

[11] Fu Y., Liu C. A., Zhu C., et al. High-performance NiO/g-C$_3$N$_4$ composites for visible-light-driven photocatalytic overall water splitting[J]. Inorganic Chemistry Frontiers, 2018, 5 (7): 1646-1652.

[12] Qin Y., Wang G., Wang Y. Study on the photocatalytic property of La-doped CoO/SrTiO$_3$ for water decomposition to hydrogen[J]. Catalysis Communications, 2007, 8 (6): 926-930.

[13] Zhang J., Yu Z., Gao Z., et al. Porous TiO$_2$ nanotubes with spatially separated platinum and CoO$_x$ cocatalysts produced by atomic layer deposition for photocatalytic hydrogen production[J]. Angewandte Chemie International Edition, 2017, 56 (3): 816-820.

[14] Liu X., Zhang Q., Liang L.,et al. In-situ growing of CoO nanoparticles on g-C$_3$N$_4$ composites with highly improved photocatalytic activity for hydrogen evolution[J]. Royal Society Open Science, 2019, 6 (7): 190433.

[15] Shi W., Guo F., Wang H., et al. New Insight of water-splitting photocatalyst: H$_2$O$_2$-resistance poisoning and photothermal deactivation in sub-micrometer CoO octahedrons[J]. ACS Applied Materials & Interfaces, 2017, 9 (24): 20585-20593.

[16] Pan C., Xu J., Wang Y., et al. Dramatic activity of C$_3$N$_4$/BiPO$_4$ photocatalyst with core/shell structure formed by self-assembly[J]. Advanced Functional Materials, 2012, 22 (7): 1518-1524.

[17] Mahendran N., Udayakumar S., Praveen K. pH-controlled photocatalytic abatement of RhB by an FeWO$_4$/BiPO$_4$ p-n heterojunction under visible light irradiation[J].New Journal of Chemistry, 2019, 43 (44): 17241-17250.

[18] Lv H., Wu X., Liu Y., et al. Photoreactivity and mechanism of BiPO$_4$/WO$_3$ heterojunction photocatalysts under simulant sunlight irradiation[J]. Ceramics International, 2018,

44 (6): 6786-6790.

[19] Pan B., Wang Y., Liang Y., et al. Nanocomposite of $BiPO_4$ and reduced graphene oxide as an efficient photocatalyst for hydrogen evolution[J]. International Journal of Hydrogen Energy, 2014, 39 (25): 13527-13533.

[20] Zhuang H., Cai Z., Xu W., et al. In situ construction of $WO_3/g-C_3N_4$ composite photocatalyst with 2D-2D heterostructure for enhanced visible light photocatalytic performance[J].New Journal of Chemistry, 2019, 43 (44): 17416-17422.

[21] Lin H., Ye H., Xu B.,et al. Ag_3PO_4 quantum dot sensitized $BiPO_4$: A novel p-n junction $Ag_3PO_4/BiPO_4$ with enhanced visible-light photocatalytic activity[J]. Catalysis Communications, 2013, 37: 55-59.

[22] Gao D., Xu Q., Zhang J., et al. Defect-related ferromagnetism in ultrathin metal-free $g-C_3N_4$ nanosheets[J]. Nanoscale, 2014, 6 (5): 2577-2581.

[23] Liu C., Jing L., He L., et al. Phosphate-modified graphitic C_3N_4 as efficient photocatalyst for degrading colorless pollutants by promoting O_2 adsorption[J]. Chemical Communications, 2014, 50 (16): 1999-2001.

[24] Wang S., Li C., Wang T.,et al. Controllable synthesis of nanotube-type graphitic C_3N_4 and their visible-light photocatalytic and fluorescent properties[J].Journal of Materials Chemistry A, 2014, 2 (9): 2885-2890.

[25] Mao Z., Chen J., Yang Y., et al. Novel $g-C_3N_4/CoO$ nanocomposites with significantly enhanced visible-light photocatalytic activity for H_2 evolution[J]. ACS Applied Materials & Interfaces, 2017, 9 (14): 12427-12435.

[26] Guo F., Shi W., Zhu C., et al. CoO and $g-C_3N_4$ complement each other for highly efficient overall water splitting under visible light[J]. Applied Catalysis B: Environmental, 2018, 226: 412-420.

[27] Peng C., Chen B., Qin Y., et al. Facile ultrasonic synthesis of CoO quantum dot/graphene nanosheet composites with high lithium storage capacity[J]. ACS Nano, 2012, 6 (2): 1074-1081.

[28] Kumar S., Baruah A., Tonda S.,et al. Cost-effective and eco-friendly synthesis of novel and stable N-doped $ZnO/g-C_3N_4$ core-shell nanoplates with excellent visible-light responsive photocatalysis[J]. Nanoscale, 2014, 6 (9): 4830-4842.

Chapter 8

Atomic Co-N$_4$ sites in 2D polymeric carbon nitride

8.1 Background

Numerous catalysts have been investigated in search of achieving optimal photocatalytic H$_2$ harvesting from water splitting. Among the various photocatalysts explored so far, polymeric carbon nitride (PCN) has received special attention owing to its suitable band gap and high thermal stability[1-3]. Nevertheless, pure PCN exhibits poor photocatalytic H$_2$ evolution efficiency because of the presence of weak van der Waals forces between the adjacent layers and the intrinsically π-conjugated system, which hamper charge transfer and cause rapid recombination of the photoexcited charge carriers and their efficient utilization[4]. Metal doping into the PCN matrix to construct M-PCN heterojunctions has been proved to be effective not only for improving charge separation but also for prominently altering the electronic and optical properties of PCN by adjusting the intrinsic electronic state and promoting charge separation, thus enhancing photoelectron transport and boosting H$_2$ generation[5,6]. Therefore, it is highly desirable to explore new M-PCN photocatalysts, especially based on cheap transition metals, to make photocatalytic H$_2$ production technology practically feasible.

Single atom catalysts refer to a special class of heterogeneous catalysts where active metal atoms are spatially isolated from each other by being immobilized, entrapped, or directly bonded to support atoms[7-12]. In the last decade, significant breakthroughs have been achieved in the synthesis of SACs for applications in thermo, photo and electro catalysis with excellent activity, selectivity and stability[13-18]. In particular, 2D materials decorated single atom catalysts have attracted immense scientific interest in order to unify and collectively utilize the distinct catalytic properties of these two classes of materials[19,20]. Among 2D materials, 2D polymeric carbon nitride (2DPCN) with abundant nitrogen sites, which can act as a Lewis acid to anchor and stabilize single atoms via metal-nitrogen bonds (M-N$_x$, M—metal atom), is considered as a promising low-cost substrate to stabilize single metal atoms[21-24]. It has been reported that atomic M-N$_x$ sites in carbon nitride can significantly improve H$_2$ harvesting activity by promoting the transfer of photoexcited electrons from CN and suppressing the recombination of e$^-$/h$^+$ pairs[25-27]. Therefore, construction of atomic M-N$_x$ sites is of great interest to explore their potential for photocatalytic hydrogen production.

Herein, in this book, we communicate the construction of atomic Co$_1$-N$_4$ sites on melamine-derived 2D polymeric carbon nitride support (Co^1@2DPCN) as a highly efficient photocatalyst for H$_2$ evoluation under visible light. The prepared Co^1@2DPCN photocatalyst was characterized by state-of-the-art tools to fully unveil the morphology, atomic dispersion, coordination environment, optical, and photo-electrochemical properties. Experimental results revealed that the formation of atomic Co$_1$-N$_4$ significantly boosted the photocatalytic efficiency of 2DPCN for H$_2$ generation from water splitting under visible light, which is also validated by performing density functional theory calculations.

8.2 Preparation of Atomic Co-N$_4$ sites in 2D polymeric carbon nitride

Melamine (C$_3$H$_6$N$_6$, 99%) was purchased from Sigma-Aldrich. Cobalt acetylacetonate dehydrate (C$_{10}$H$_{14}$CoO$_4$·2H$_2$O) was obtained from Alfa Aesar. Ethanol (C$_2$H$_6$O, 70%) and methanol (CH$_4$O, 90%) were bought from SAMCHUN Chemicals.

Bulk 2D dimensional polymeric carbon nitride (2DPCN) was synthesized according to a reported method with slight modifications. In a typical procedure melamine (5g, 39.645 mmol) was placed in a porcelain boat with a cover and subsequently heated at 550 °C in a muffle furnace for 2 h at a rate of 2 °C/min to obtain bulk CN, which was milled to powder form with a mortar for further process. Next, 2g of bulk 2D polymeric carbon was spread in a porcelain boat with a lid and treated in a tube furnace under a N$_2$ atmosphere at 560 °C at a ramping rate of 2 °C/min for 2 h for exfoliation of bulk 2DPCN into thin 2D polymeric carbon nitride sheets. After cooling to room temperature, the obtained yellow colored product was collected and sealed in a vial for further use.

Fabrication of atomic cobalt on 2D polymeric carbon nitride was carried out by wet impregnation method involving two steps. First 0.5 g of the as prepared 2D polymeric carbon nitride support was dispersed in 10ml of ethanol under sonication for 5 minutes, followed by the addition of 0.8wt% Cobalt acetylacetonate dihydrate (20.1 mg) and further sonication for five minutes minute. After that, the resultant mixture was stirred at 60 °C for uniform impregnation of cobalt precursor on 2DPCN and to evaporate the ethanol.   Finally, the yellow colored product was tempered at 400 °C in a tube furnace under 5% H$_2$ at a ramping rate of 2.5 °C/min to reach the target temperature and kept at the same temperature for 2hrs. After cooling to room temperature, the product was designated as Co^1@2DPCN and kept in an air tight glass vial for further characterization and catalysis.

8.3   Characterization of Co^1@2DPCN

8.3.1   XRD of Co^1@2DPCN

Powder XRD measurements were executed on a Rigaku RU-200b X-ray powder diffractometer with Cu Kα radiation (λ = 1.5406 Å) operated at 45 kV and an emission current of 50 mA. The process for constructing atomic Co$_1$-N$_4$ sites into 2D polymeric carbon nitride (2DPCN) is schematically shown in Figure 8.1. 2DPCN support was prepared in two steps by the thermal polymerization of melamine monomers first in air then under N$_2$ and later confirmed by SEM, TEM, XRD and FTIR analysis. The XRD pattern of 2DPCN showed two prominent peaks at 13° and 27.4° (Figure 8.2), indicating the formation of the carbon nitride structure[28]. The intense peak at 27.4° corresponds to the (002) plane with an interlayer spacing of 0.326 nm, suggesting the formation of a multilayer graphite-like structure. The weak diffraction peak at 13° is ascribed to the repeated in-plane (001) arrangement of the tri-s-triazine structural units[29]. Weakening of the diffraction peaks' intensity after treating bulk CN under N$_2$ reveals the effective exfoliation of the bulk carbon nitride into the thinner 2DPCN structure. Field emission scanning electron microscopy (FE-SEM) and TEM images   also unveiled the formation of thin 2D sheet-like morphology for the as synthesized polymeric carbon nitride (Figure 8.3).

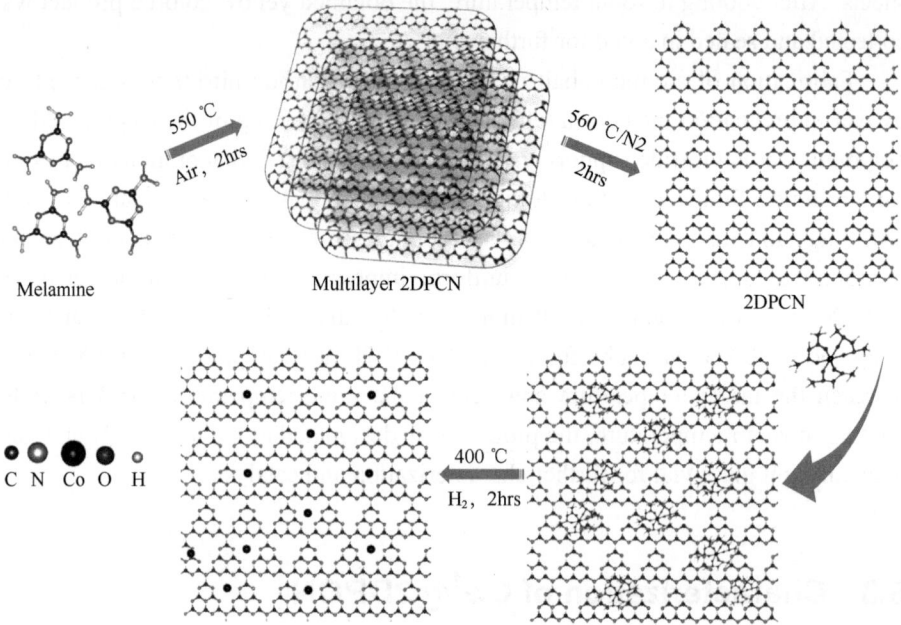

Figure 8.1   Graphic illustration of different steps executed for the synthesis of Co^1@2DPCN photocatalyst

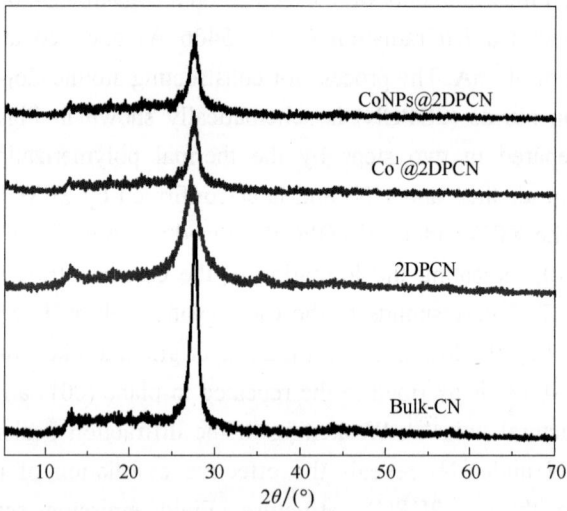

Figure 8.2   XRD patterns of the as prepared polymeric carbon nitride along with Co^1@2DPCN and CoNPs@2DPCN

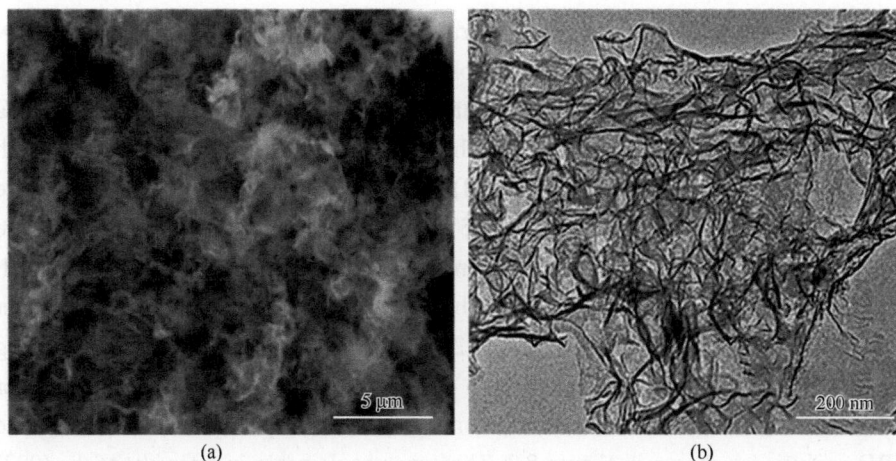

(a)   (b)

Figure 8.3   Field emission scanning electron microscopy (FE-SEM) and TEM images of 2DPCN

8.3.2   FT-IR, BET, SEM and TEM of Co^1@2DPCN

The morphology of the pure 2DPCN as well as Co^1@2DPCN samples was analyzed on HITACHI SU8000 field emission scanning electron micro-scope (FE-SEM) instrument operated at an accelerating voltage of 10.0 kV. TEM images were obtained on a JEOL JEM-2100 (RH) machine with an operational accelerating voltage of 200 kV. The FTIR spectral analysis was recorded with a Thermo Mattson, Infinity Gold Fourier transform infrared spectrometer provided with a liquid nitrogen-cooled narrow band MCT detector, using an attenuated total reflection cell equipped with a Ge crystal. The surface area was calculated by the multi-point Brunauer-Emmett-Teller(BET) theory, while pore size distribution was estimated from the adsorption data by applying BJH model.

As shown in Figure 8.4, FT-IR spectra of prepared 2DPCN exhibited typical features of graphitic carbon nitride structure. A strong peak around 806 cm^{-1} corresponds to the bending vibrations of the aromatic C-N bond in the outer plane and an adjacent peak at 890 cm^{-1} is attributed to the deformation mode of the cross-linked heptazine units. The peaks in the 1200-1700 cm^{-1} region are ascribed to the vibrational modes of the conjugated aromatic C-N frameworks. The absorption bands at 3160 cm^{-1} and 3430 cm^{-1} are accredited to the stretching and vibrational modes of the O-H and N-H bonds, respectively[30,31]. A new absorption band at appeared 2182 cm^{-1} in the FTIR spectrum of 2DPCN after N$_2$ treatment belongs to the asymmetric stretching vibration of C≡N bond. The formation of C≡N bond can promote electron delocalization and modify band structures to improve visible-light absorption and charge separation. N_2 adsorption isotherm revealed a high surface area of 30.42 m²/g for the 2DPCN, and a porous structure with cumulative Barrett-Joyner-Halenda (BJH) pore volume of 0.239 cm³/g and average pore width of 156.202 Å (Figure 8.5). Subsequently, Co single atoms were decorated on 2DPCN by impregnating and adsorbing the desired amount of Cobalt acetylacetonate dihydrate on 2DPCN in ethanol, followed by drying and treating under H_2 atmosphere at 400 °C to obtain Co¹@2DPCN (details are given in the proceeding experimental section). Apart from the two characteristic diffraction peaks observed for pure 2DPCN at 27.4° and 13.0°, no additional peaks for Co species were detected in the XRD analysis, as shown in Figure 8.2. The apparent decrease in intensity of the (002) peak indicates more distorted stacking arrangement of the carbon nitride layers and formation of coordination between Co and N atoms of the 2DPCN host, which is in good agreement with other M-PCN materials[32-34]. As shown in Figure 8.5, no significant difference was observed in the IR spectra of the pure and Co doped 2DPCN, suggesting that Co induction produced no changes in the skeleton structure of 2DPCN. SEM and TEM analysis revealed that 2DPCN retained its sheet-like structure after Co induction and subsequent pyrolysis treatment (Figure 8.6).

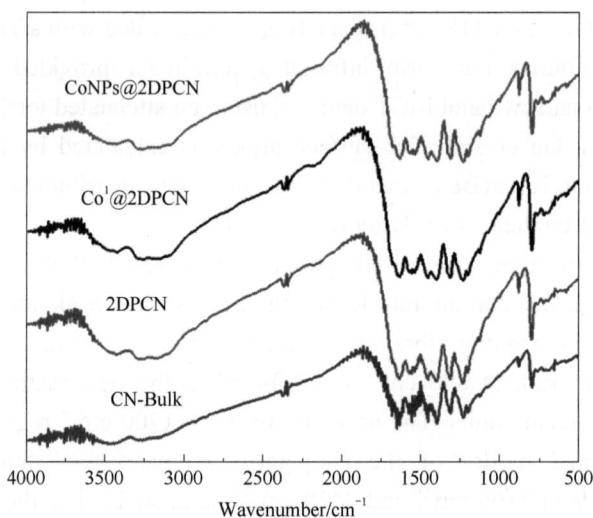

Figure 8.4   FT-IR patterns of the as prepared polymeric carbon nitride along with Co¹@2DPCN and CoNPs@2DPCN

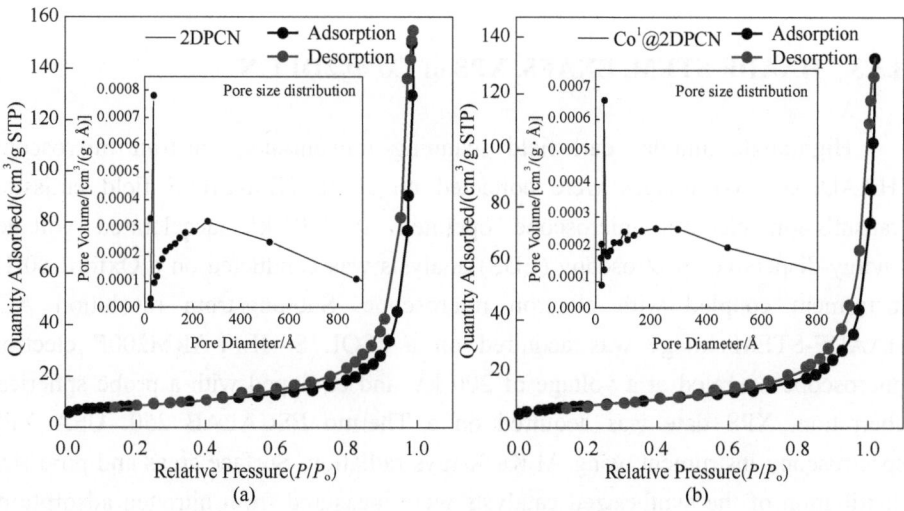

Figure 8.5　Nitrogen adsorption−desorption isotherms and BJH pore size distribution curves for
(a) 2DPCN and (b) Co¹@2DPCN

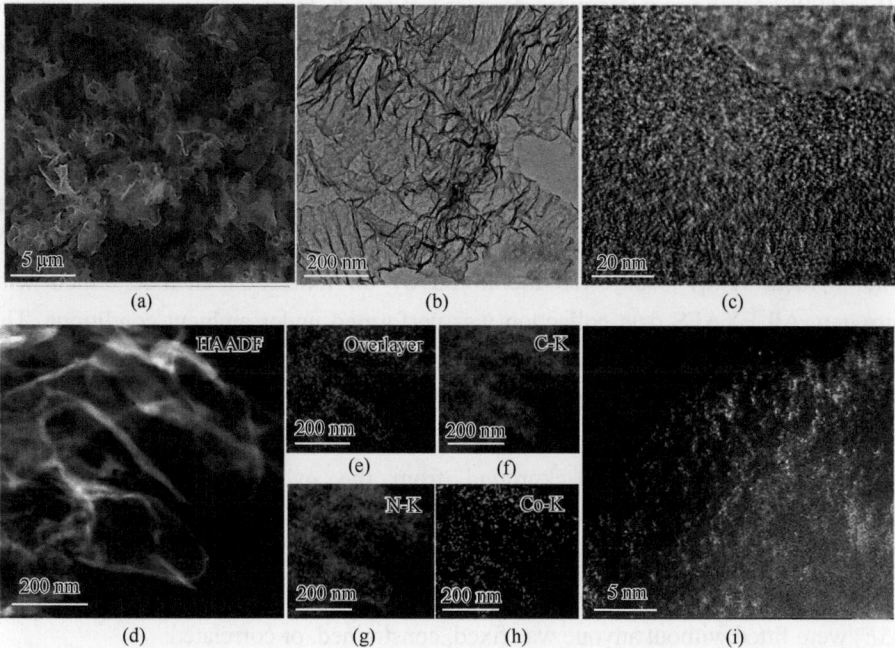

Figure 8.6　Electron microscopy characterization of Co¹@2DPCN. (a) SEM image, (b) TEM
image, (c) high magnification HR-TEM image, (d) The high-angle annular dark-fieldscanning
TEM (HAADF-STEM) image of Co¹@2DPCNHR-TEM image, (e)-(h) EDX elemental mapping
for C,N, and Co and (j) Abbreviation corrected HAADF-STEM image (bright dots corresponds to
Co single atoms)

8.3.3 HAADF-STEM, EXAFS, XPS of Co^1@2DPCN

High-angle annular dark-field scanning transmission electron microscopy (HAADF-STEM) images were collected on JEOL JEM-2100F field emission transmission electron microscope operated at 200 kV accelerated voltage. Energy-dispersive spectroscopy (EDS) analysis was conducted on a Oxford 80HT instrument coupled with electron microscope. Sub-angstrom resolution AC-HAADF-STEM image was acquired on a JEOL S-4JEM-ARM200F electron microscope operated at a voltage of 200 kV and equipped with a probe spherical aberration. XPS date was acquired on a Thermo ESCALAB 250, USA XPS spectroscopy instrument using Al Ka X-rays radiation. Surface areas and pore size distribution of the synthesized catalysts were measured from nitrogen adsorption-desorption isotherms performed on a NOVA4200e nitrogen adsorption analyzer (Quanta chrome Instruments, USA) at 77 K under a relative pressure p/p_0 of 3.3 × 10^{-7}–0.995. All the samples were degassed at 574 K for 2 h under vacuum prior to analysis for the sake of removing adsorbed gas or moisture. The X-ray absorption fine structure spectra (EXAFS) data were collected at the 1W1B station in the Beijing Synchrotron Radiation Facility (BSRF) at operational voltage of 2.5 GeV with maximum current of 250 mA. XAFS data for atomic Co-K was collected under fluorescence excitation mode, while that of references (Co foil and Cobalt oxide) were acquired under transmission mode using Si (111) double-crystal monochromator. The incident X-ray beam was monitored by an ionization chamber filled with nitrogen. All EXAFS data collection was performed under ambient conditions. The acquired EXAFS data was treated according to the standard procedures using the ATHENA module implemented in the IFEFFIT software packages in the range k of 12-10 Å^{-1} and R range of 1-3 Å. The k^3-weighted EXAFS spectra were obtained by subtracting the post-edge background from the overall absorption and then normalized with respect to the edge-jump step. Subsequently, the k^3- weighted $\chi(k)$ data of the Co K-edge were FT to real (R) space using hanging. The four parameters, coordination number (C.N), bond length (R), Debye-Waller factor (σ^2) and E^0 shift (ΔE) were fitted without anyone was fixed, constrained, or correlated.

High resolution TEM and HAADF-STEM analysis of Co^1@2DPCN excluded the formation of any Co nano clusters or particles and indicated a high dispersion for Co on 2DPCN probably in the atomic state in Figure 8.7 (c, d). Energy dispersive X-ray mapping (EDX) revealed the presence and homogeneous distribution of Co on the 2DPCN support in Figure 8.7 (e)-(h). The actual presence of Co atoms on 2DPCN was demonstrated by performing aberration-corrected HAADF-STEM analysis, where individual Co atoms appeared as bright spots due to the high Z-contrast between the cobalt and the 2DCN support as shown in Figure 8.7 (i). The induction of atomic Co leads to an increase in the BET surface area of 2DPCN to 34.6818 m²/g with cumulative pore volume of 0.203 cm³/g, while average pore size decreases to 132.913Å (Figure 8.6). In contrast, TEM and HR-TEM image of CoNPS@2DPCN clearly revealed the formation of Co nanoparticles (Figure 8.8).

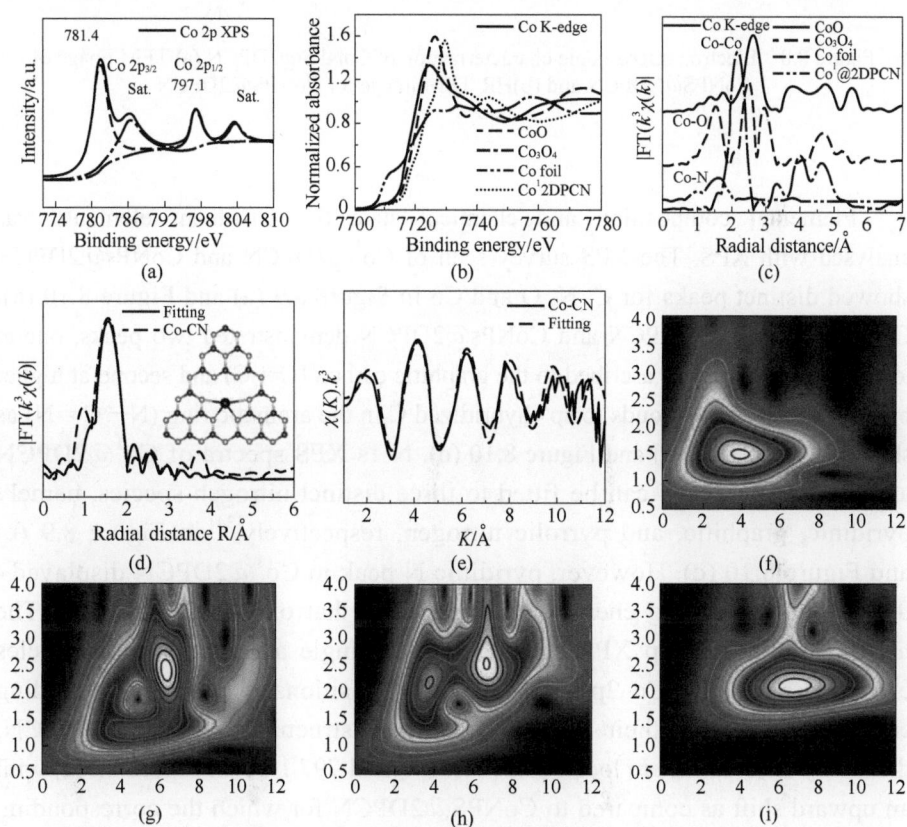

Figure 8.7   XPS and XAFS analysis

(a) High-resolution XPS Co 2p spectra of Co¹@2DPCN

(b) Normalized Co K-edge XANES

(c) K-edge FT-EXAFS in R space for Co¹@2DPCN and Co-foil, CoO and Co₃O₄ references

(d)-(e) EXAFS fitting curves in R space and k space

(f)-(i) Wavelet transforms for the k³-weighted EXAFS signals for Co¹@2DPCN, Co₃O₄, CoO and Co-foil respectively

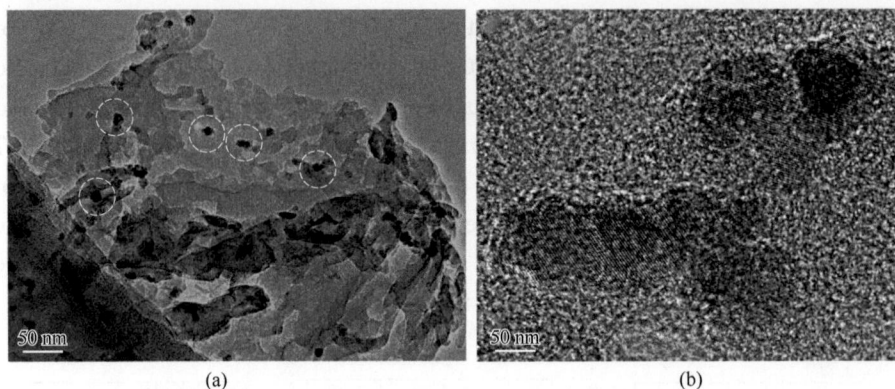

(a)   (b)

Figure 8.8   Electron microscopic characterization of CoNPS@2DPCN (a) TEM image of
CoNPS@2DPCN and (b)HR-TEM image for CoNPS@2DPCN

Elemental composition and chemical state of constituent elements was analysed with XPS. The XPS survey scan of Co^1@2DPCN and CoNPs@2DPCN showed distinct peaks for C, N, O and Co in Figure 8.9 (a) and Figure 8.10 (a). C 1s XPS of Co^1@2DPCN and CoNPs@2DPCN demonstrated two peaks, one at lower binding energy is ascribed to the graphitic carbon (C=C) and second at higher binding energy corresponds to sp^2-hybridized C in the aromatic ring (N—C=N) as shown in Figure 8.9 (b) and Figure 8.10 (d). N 1s XPS spectra of Co^1@2DPCN and CoNPs@2DPCN can be fitted to three distinct nitrogen species, namely pyridinic, graphitic, and pyrrolic nitrogen, respectively[34] in Figure 8.9 (c) and Figure 8.10 (c). However, pyridinic N peak in Co^1@2DPCN displayed~ 0.5 eV higher binding energy as compared to that of CoNPs@2DPCN. The high resolution Co 2p XPS spectra both of single atoms and nanoparticles exhibited peaks for Co $2p_{1/2}$ and Co$2p_{3/2}$ transitions with different binding energies, suggesting their different electronic structures. For Co^1@2DPCN, the Co 2p XPS peaks are located at 781.6 eV and 797.1 eV in Figure 8.7(a) with an upward shift as compared to CoNPS@2DPCN for which the corresponding peaks are centred at 780.4 eV and 796.6 eV in Figure 8.10 (b). The Co 2p XPS binding energies of CoNPs are closer to Co^{3+} (780.1 eV and 795.4 eV) while that of Co single atom are closer to Co^{2+} (782.8 eV and 798.6 eV)[35]. From XPS it can be concluded that Co atoms are strongly coordinated with pyridinic N of 2DPCN to construct Co-N heteroatom junction, which is resulted from electron donation from nitrogen to cobalt as reported in literature[36,37].

128

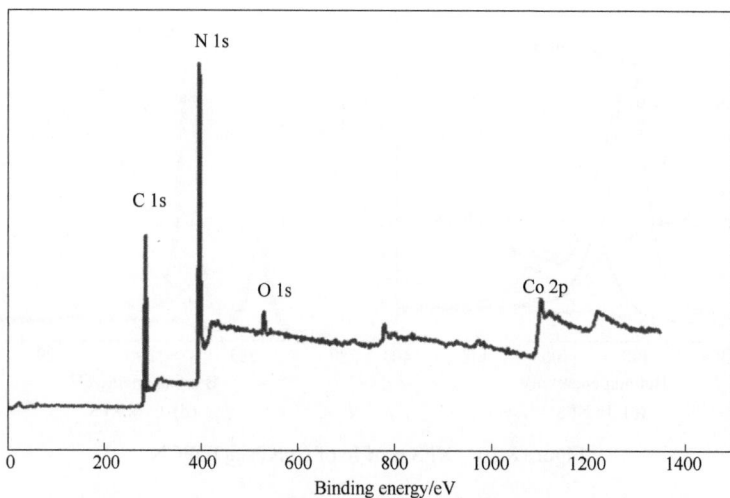

(a) XPS survey scan for Co¹@2DPCN

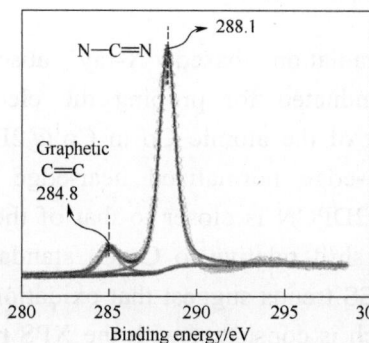

(b) C 1s XPS

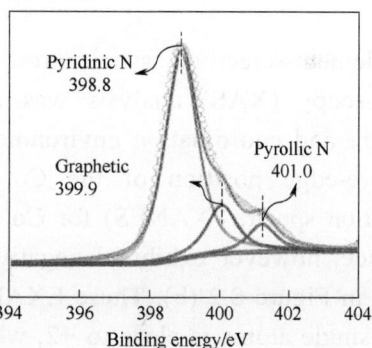

(c) N 1s XPS

Figure 8.9　XPS analyses of Co¹@2DPCN

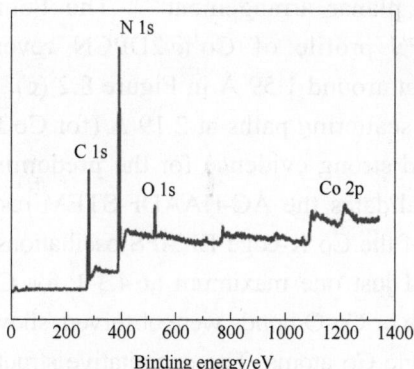

(a) XPS survey scan for CoNPs@2DPCN

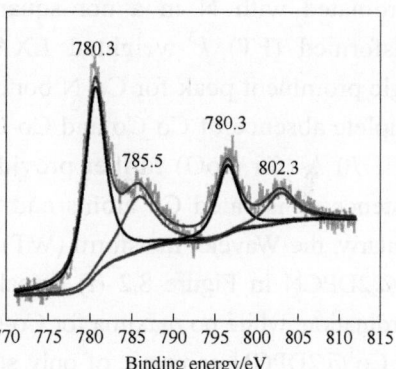

(b) Co 2p XPS

Figure 8.10

129

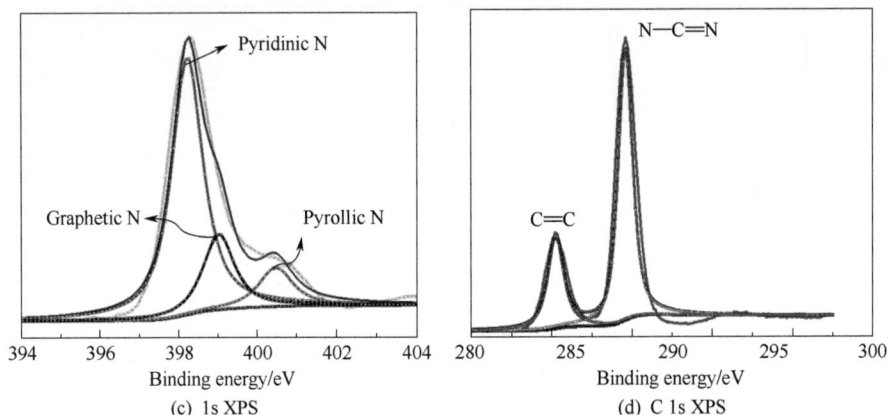

Figure 8.10　XPS profile of CoNPs@2DPCN

Element-selective synchrotron radiation based X-ray absorption spectroscopy (XAS) analysis was conducted for probing the electronic structure and coordination environment of the atomic Co in Co^1@2DPCN. The pre-edge position of the Co K-edge normalized near-edge X-ray absorption spectra (XANES) for Co^1@2DPCN is closer to that of the CoO reference, however exhibited negative shift relative to Co_3O_4 standards as shown in Figure 8.2 (b). These EXANES trends suggest that oxidation state of Co single atoms is close to +2, which is consistent with the XPS results. The absence of pre-edge peak at about 7715 eV, which corresponds to the square-planar structure of Co-N_4, suggests that Co atoms in 2DPCN are coordinated with N in a non-square planar arrangement[38]. The Fourier-transformed (FT) k^3-weighted EXAFS profile of Co^1@2DPCN revealed single prominent peak for Co-N bond at around 1.59 Å in Figure 8.2 (c). The complete absence of Co-Co and Co-O scattering paths at 2.19 Å (for Co foil) and 1.70 Å (for CoO) further provided strong evidence for the predominant existence of isolated Co atoms and validates the AC-HAADF-STEM results. Similarly, the Wavelet transform (WT) of the Co K-edge EXAFS oscillations for Co^1@2DPCN in Figure 8.2 (f) revealed just one maximum at 4.3-1 for Co-N coordination, while no maxima for Co-Co or Co-O bonds were observed, showing that Co^1@2DPCN comprise of only single Co atoms. The quantitative structural parameters for atomic Co sites in Co^1@2DPCN were acquired by performing least squares EXAFS curve fitting analysis. The fitting curve overlapped well with the experimental data as shown in Figure 8.3 (d). According to the fitting parameters , each Co atom is coordinated to four nitrogen atoms with mean Co–N bond length of 2.05Å.

8.3.4   Optical and electrochemical properties assessment

It is well known that the photocatalytic hydrogen evolution efficiency of semiconductors highly depends on their optical and electrochemical properties[39-40]. Optical and electrochemical impedance spectroscopy, along with Mott-Schottky analysis, was used to investigate the effect of Co doping on the optical and photoelectric properties of the 2DPCN. As shown in Figure 8.11 (a), compared to pure 2DPCN and CoNPs@2DPCN, UV-Visible diffuse reflectance spectra of Co^1@2DPCN exhibited extended absorption tail in the visible region, inferring the improved visible light harvesting efficiency and enhanced photoexcited electron-hole capturing ability of Co^1@2DPC. This phenomenon is credited to the strong absorption efficiency of Co in the UV-Visible light region[41,42]. As illustrated in Figure 8.11 (b), atomic Co doping significantly reduced the band gap energy (Eg) for Co^1@2DPCN (2.63 eV) when compared to CoNPs@2DCN (2.76 eV) and 2PDCN (2.86 eV). These features indicate alteration in the electronic band structure of 2DPCN after Co doping.   Charge separation and transfer efficiency of as prepared catalysts was evaluated with photoluminescence (PL) emission spectroscopy and electrochemical impedance spectroscopy (EIS). The intensity and position of PL emission spectra strongly correlates with the recombination rate of photoexcited electrons-holes pairs. In comparison to pure 2DCN and CoNPs@2DPCN, Co^1@2DPCN PL emission spectra showed a red shift with lower intensity in Figure 8.11 (c), indicating better charge separation and faster interfacial charge transfer ability of Co^1@2DPCN than 2DCN and CoNPs@2DPCN. Greater charge separation and low charge transfer resistance for Co^1@2DPCN is further validated by the smaller arc radius of Nyquist plot than the other catalysts [Figure 8.11 (d)]. Photoelectrochemical analysis revealed higher current density for Co^1@2DPCN than that of 2DPCN and CoNPs@2DPCN, indicating substantially improved charge separation efficiency after photoactivation in Figure 8.11 (e). Besides, the band gap appropriate band structure also plays a critical role for optimizing the photocatalytic activity. Accordingly, as shown in Figure 8.11 (f), flat band potential derived from the linear x-intercept region of the Mott–Schottky (M-S) plots of 2PDCN, $Co^1@2PDCN$ and CoNps@2PDCN was estimated to be −0.49 eV, −0.44 eV and −0.36 eV (vs. SCE) respectively. Consequently, by applying Nernst equation, conduction band potential (E_{CB}) for 2PDCN, $Co^1@2PDCN$ and CoNps@2PDCN was determined to be −0.25 eV, −0.20 eV and −0.12 eV (versus NHE), respectively. The valence band potential (E_{VB}) of 2DPCN, CoNP@2DPCN and $Co^1@2DPCN$ calculated from their band gap energies was found to be 2.61, 2.58 and 2.51 eV, respectively.

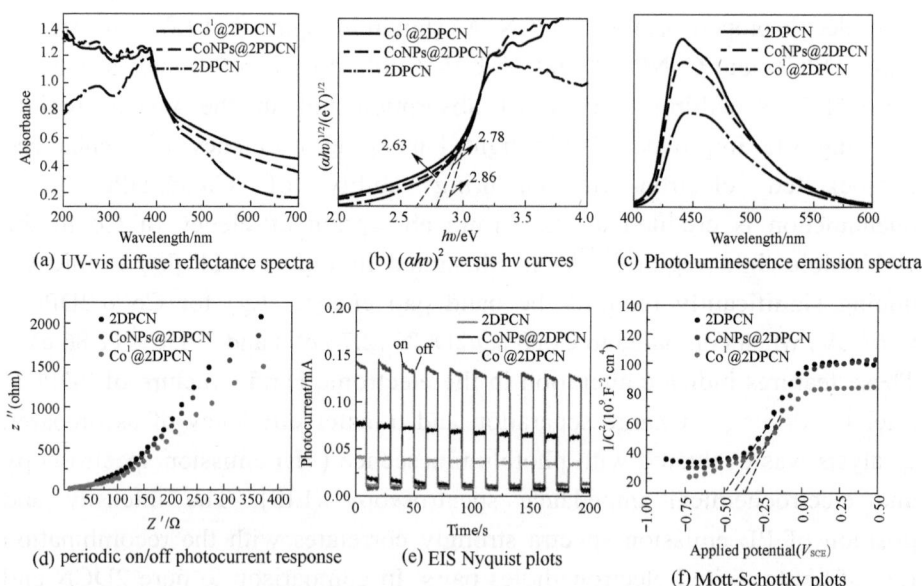

(a) UV-vis diffuse reflectance spectra (b) $(ahv)^2$ versus hv curves (c) Photoluminescence emission spectra

(d) periodic on/off photocurrent response (e) EIS Nyquist plots (f) Mott-Schottky plots

Figure 8.11 Optical and electrochemical analysis of as prepared catalysts

8.4 Photocatalytic performance

8.4.1 Photocatalytic H$_2$ generation testing

Given the excellent photoelectric properties, we investigated $Co^1@2DCN$ for photocatalytic hydrogen generation from water splitting and compared its activity to that of pure 2DPCN and CoNPs@2DPCN as shown in Figure 8.12 (a).

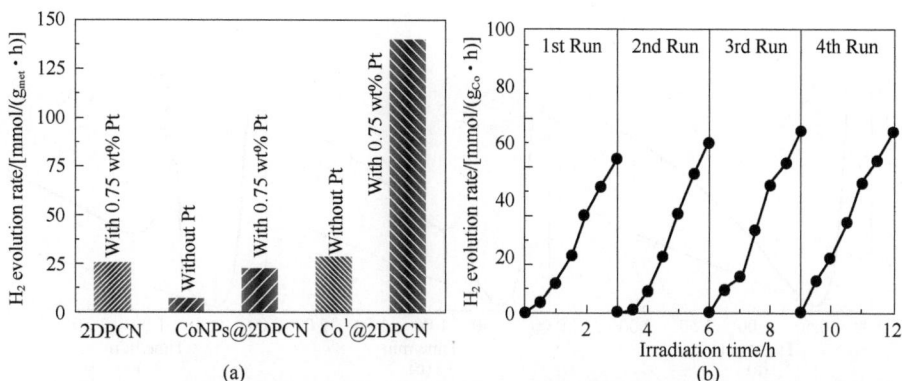

Figure 8.12   (a) Photocatalytic performance of pure 2DPCN, CoNPs@2DPCN and Co¹@2DCPN and (b) stability test of Co¹@2DCPN

For Photolysis and experiments conducted in the dark, no gas evolution was detected. Pure 2PCN catalyst demonstrated very poor photocatalytic H_2 evolution activity and yielded only a trace amount of hydrogen which is possibly resulted from the rapid recombination of photoexcited electron-hole pairs. However, when 0.75 wt% of Pt was employed as co-catalyst 2DPCN displayed significantly enhanced photocatalytic activity with an average H_2 evolution rate of 25.3 $mmol_{H_2}/(h \cdot g_{Pt})$. Notably, Co doping into the 2DPCN matrix produced pronounced effect on the photocatalytic H_2 harvesting efficiency even without using Pt as co-catalyst. CoNPs@2DPCN produced hydrogen at a rate of 6.8 $mmol_{H_2}/(h \cdot g_{Co})$ and 22.4 $mmol_{H_2}/(h \cdot g_{metal})$ (0.75 wt% Pt). As expected , atomic Co¹@2DCN exhibited optimal photocatalytic activity with a mass-normalize H_2 production rate of 28.3 $mmol_{H_2}/(h \cdot g_{Co})$, which is ~4.2 times higher than that of CoNPs. Surprisingly, single atom Co¹@2DCN with same Co content (0.8 wt%) demonstrated better H_2 harvesting efficiency as that of Pt (0.75 wt%) as co-catalyst, which clearly manifests the crucial role of the atomic Co¹-N₄ sites for photocatalytic H_2 harvesting. More importantly, our as constructed Co₁-N₄ atomic sites exhibited superior H_2 evolution rate as compared to the previously reported Co₁-N₄ sites [51]. Remarkably, with 0.75 wt% Pt hydrogen evolution rate for Co¹@2DPCN reached as high as 138.6 $mmol_{H_2}/(h \cdot g_{metal})$, which is ~5.5 and ~6.2 times higher as compared to 2DPCN and CoNPs@2DPCN respectively. The AQE for Co¹@2DPCN measured at 420 nm was estimated to be 1.38%.

Other characterizations are shown in Figure 8.13-Figure 8.15.

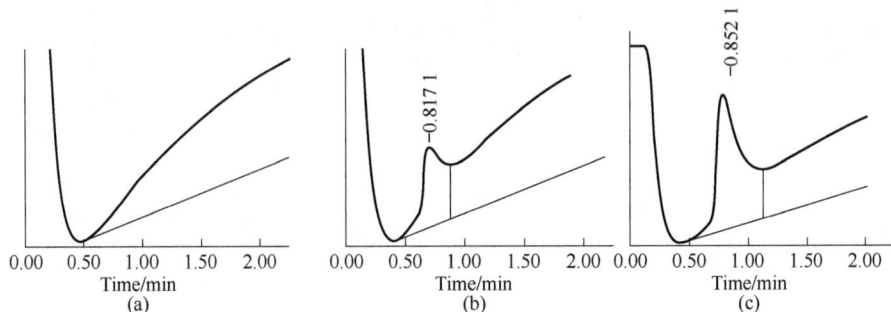

Figure 8.13 GC signals of photocatalytic H_2 evolution in TEOA aqueous solution for (a) 2DPCN, (b) CoNPS@2DPCN and (c) Co^1@2DPCN

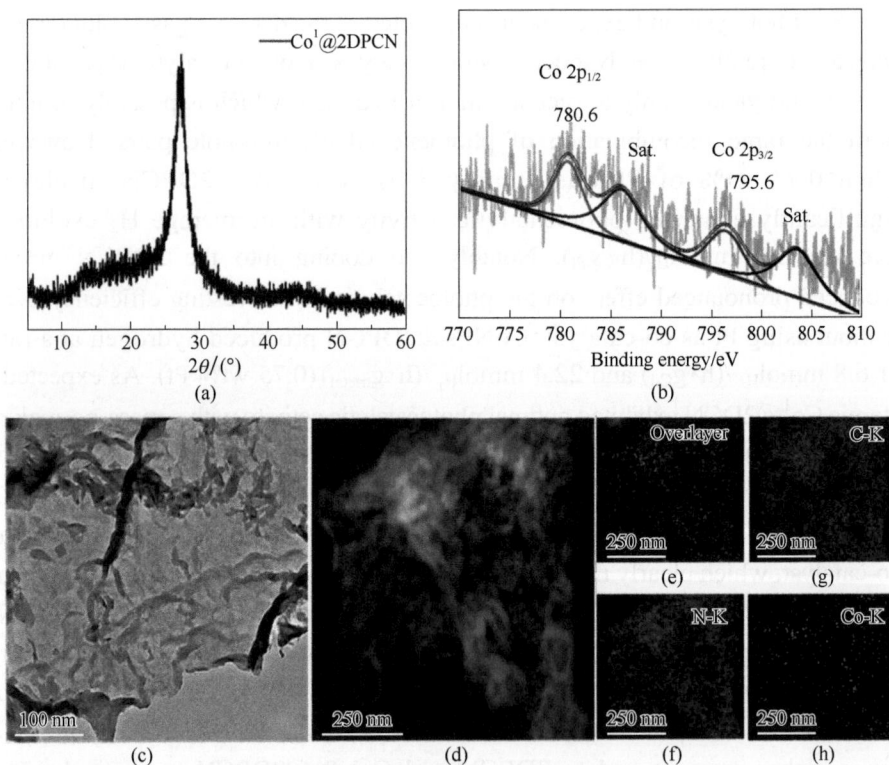

Figure 8.14 Characterization of Co^1@2DPCN catalyst after four cycles of photocatalysis (a) XRD patterns of Co^1@2DPCN, (b) TEM image of Co^1@2DPCN, (c) HAADF-STEM image of Co^1@2DPCN, (d)-(f) EDS elemental mapping of Co^1@2DPCN

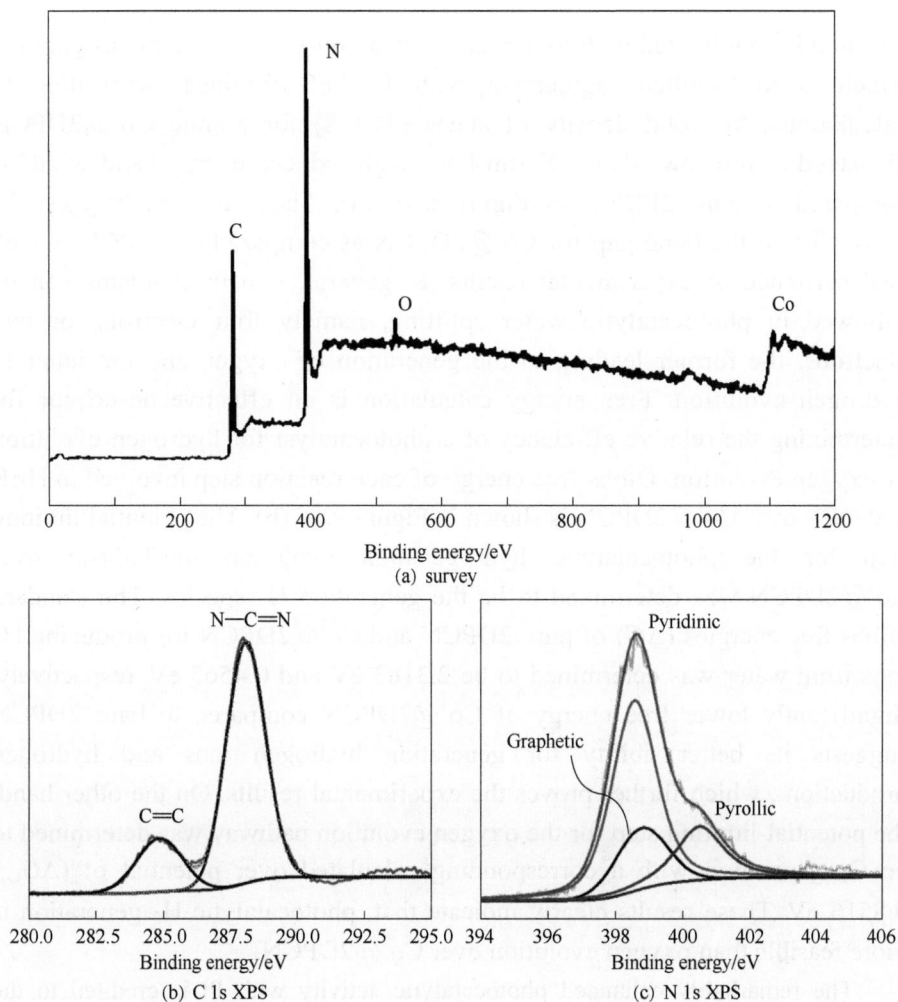

Figure 8.15　XPS Characterization of Co¹@2DPCN catalyst after four cycles of Photocatalysis

8.4.2　Density functional theory calculations

Density functional theory calculations (DFT) were further performed to simulate the change in the band structure of 2PCN after doping atomic Co^1-N_4 sites and to unveil the underlying mechanism of H_2 evolution over Co^1@2DPCN catalyst. In the optimized Co^1@2DPCN structure in Figure 8.16 (c)

Co atom is coordinated with four N atoms in a non-square planar configuration which is in excellent agreement with EXAFS findings. According to calculations, the total density of states (TDOS) for atomic Co^1@2DPCN displayed a shift towards the Fermi level with reduced energy band width as compared to pure 2PCNC in Figure 8.16 (a). These features suggest the narrowing of the band gap for Co^1@2DPCN as compared to 2DPCN, which is consistence of experimental results. In general, two mechanisms can be followed in photocatalytic water splitting, namely four electrons or two electrons, the former leading to the generation of oxygen and the latter to hydrogen evolution. Free energy calculation is an effective descriptor for determining the relative efficiency of a photocatalyst for hydrogen evolution or oxygen evolution. Gibbs free energy of each reaction step involved in HER and OER over Co^1@2DPCN is shown in Figure 8.16 (b). The potential limiting step for the photocatalytic hydrogenation evolution mechanism over Co^1@2DPCN was determined to be the generation H^+ species. The standard Gibbs free energies (ΔG) of pure 2DPCN and Co^1@2DPCN for producing H^+ ions from water was determined to be 2.3183 eV and 0.4565 eV, respectively. Significantly lower free energy of Co^1@2DPCN compared to bare 2DPCN suggests its better ability for generating hydrogen ions and hydrogen production , which further proves the experimental results. On the other hand, the potential-limiting step for the oxygen evolution pathway was determined to be O_2 generation with a corresponding calculated over potential of (ΔG_{O_2}) 0.8516 eV. These results clearly indicate that, photocatalytic H_2 generation is more feasible than oxygen evolution over Co^1@2DPCN.

The remarkably enhanced photocatalytic activity with Pt is credited to the synergistic effect between Co^1-N_4 atomic sites Pt. Pt is beneficial not only for trapping photo excited electrons but also for fast proton reduction. Meanwhile, atomic Co doping not only improved the band structure but also promoted the directional transfer of photoexcited electrons through the newly constructed Co-N_4 heteroatom junction. Thus, efficient mobility of electrons to the surface prohibited the recombination of e^-/h^+ pairs and boosted the H_2 evolution rate. Moreover, 12 hour cycling photocatalytic experiment revealed a linear increase in H_2 production, thus probing highly stable photocatalytic performance of Co^1@2DPCN in Figure 8.16(b). Post XRD, TEM, HR-TEM characterizations revealed no distortion in the sheet like the structure of 2DPCN and aggregation of Co single atoms into nanoparticles in Figure 8.16 (a), (c)-(h), suggesting the highly stability of

Co^1@2DCN. However, Co 2p XPS peaks of the used catalysts are shifted towards low binding energy in Figure 8.16 (b), suggesting the partial reduction of cobalt single atoms and probing the transfer of photoexcited electrons from 2DPCN to Co surface during Photocatalysis. Thus, Co^1@2DCN indeed shows remarkable photocatalytic H$_2$ evolution activity and stability owing to its high visible light absorbance as well as low photoexcited electron-hole pairs recombination rate.

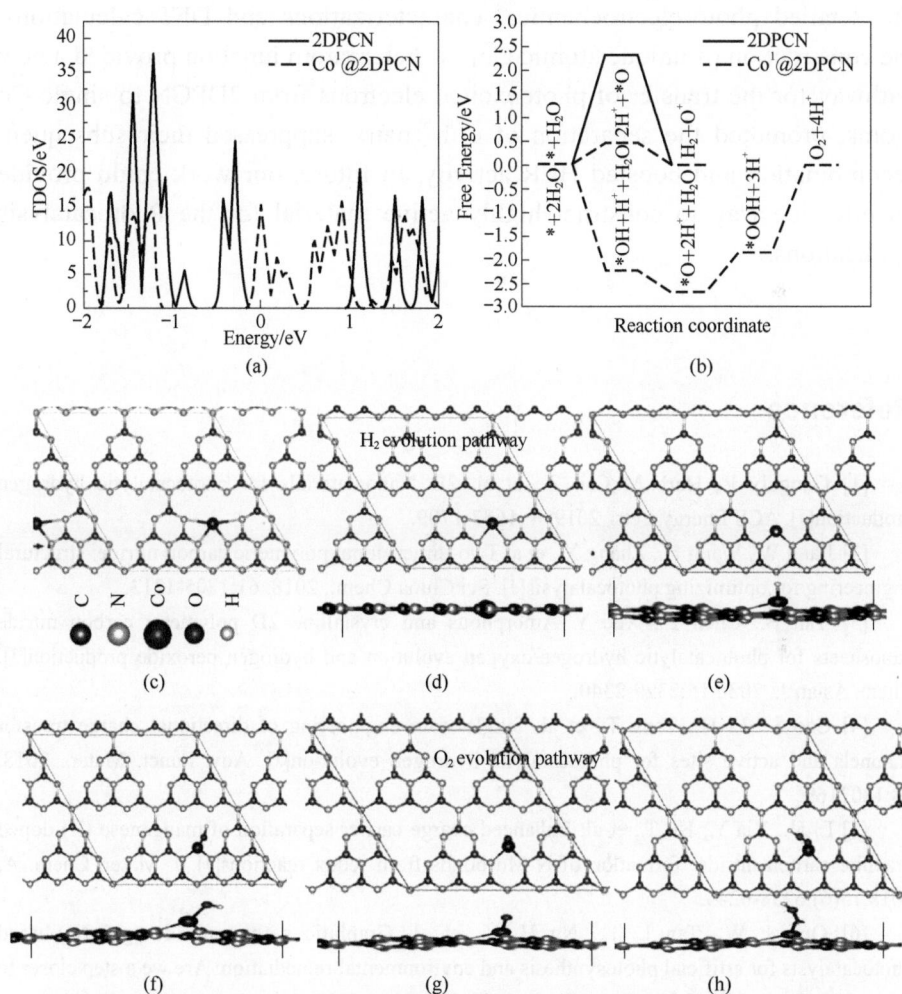

Figure 8.16   (a) Calculated total density of states for 2DPCN and Co^1@2DPCN, (b) Gibbs free energy profile for the HER and OER for 2DPCN and Co^1@2DPCN, (c) Optimized structure for Co^1@2DPCN, (d)-(e) H$_2$ evolution mechanism on Co^1@2DPCN and (f)-(h) O$_2$ evolution mechanism on Co^1@2DPCN

8.5 Conclusion

A simple two-step process has been demonstrated for the fabrication of a single atom $Co^1@2DPCN$ as a highly efficient visible light-driven H_2 harvesting photocatalyst. AC-HAADF-STEM and XAS analyses disclosed that Co single atoms were coordinated with four N atoms of 2DPCN in a non-square planar configuration to form atomic Co^1-N_4 sites. According to the detailed photo-electrochemical characterizations and DFT calculations, the construction of unique atomic Co_1-N_4 heteroatom junction provided a new pathway for the transfer of photoexcited electrons from 2DPCN to single Co atoms, promoted the separation of e^-/h^+ pairs, suppressed their subsequent recombination and boosted HER activity. In future, our work could provide an effective way to construct highly active material for the Photocatalysis applications.

Reference

[1] Ganguly P., Harb M.,Cao Z., et al. 2D Nanomaterials for photocatalytic hydrogen production[J]. ACS Energy Lett., 2019, 4: 1687-1709.

[2] Jiang W., Wang H., Zhang X., et al.Two-dimensional polymeric carbon nitride: structural engineering for optimizing photocatalysis[J]. Sci China Chem., 2018, 61:1205-1213.

[3] Yan B., Chen Z., Xuc Y. Amorphous and crystalline 2D polymeric carbon nitride nanosheets for photocatalytic hydrogen/oxygen evolution and hydrogen peroxide production[J]. Chem. Asian J., 2020,15:2329-2340.

[4] Cao S., Li H., Tong T., et al. Single-atom engineering of directional charge transfer channels and active sites for photocatalytic hydrogen evolution[J]. Adv. Funct. Mater., 2018, 28:1802169.

[5] Li H., Xia Y., Hu T., et al. Enhanced charge carrier separation of manganese (II)-doped graphite carbon nitride: formation of N-Mn bonds from redox reactions[J]. J. Mater. Chem. A, 2018,15(6):6238-6243.

[6] Ong J. W., Tan L. L., Ng H. Y., et al. Graphitic carbon nitride (g-C_3N_4)-based photocatalysts for artificial photosynthesis and environmental remediation: Are we a step closer to achieving sustainability? [J]. Chem. Rev. , 2016,116: 7159-7329.

[7] Kaiser K. S., Chen Z., Akl M. F. D., et al. Single-atom catalysts across the periodic table[J]. Chem. Rev. ,2020,120: 11703-11809.

[8] Wang A., Li J., Zhang T. Heterogeneous single- atom catalysis[J]. Nat. Rev. Chem.,

2018,2: 65-81.

[9] Li X., Huang Y., Liu B. Single-atom catalysis: directing the way toward the nature of catalysis chem[J]. Adv. Mater., 2019, 5: 2733-2739.

[10] Liu J. Catalysis by supported single metal atoms[J]. ACS Catal., 2017,7: 34-59.

[11] Xiong H., Datye K. A., Wang Y. Thermally stable single-atom heterogeneous catalysts[J]. Adv. Mater., 2021, 33: 2004319.

[12] Zhang Q., Guan J. Applications of single-atom catalysts[J]. Nano Res. ,2022,15:38-70.

[13] Li J., Li Y., Zhang T. Recent progresses in the research of single-atom catalysts[J]. Sci China Mater.,2020,63:889-891.

[14] Mitchell S., Pérez-Ramírez J. Single atom catalysis: a decade of stunning progress and the promise for a bright future[J]. Nat Commun., 2020,11: 4302.

[15] Cao C., Song W. Single-atom catalysts for thermal heterogeneous catalysis in liquid: Recent progress and future perspective[J]. ACS Materials Lett.,2020, 2:1653-1661.

[16] Datye K. A., Guo H. High temperature shockwave stabilized single atoms[J]. Nat. Commun., 2021, 12:89.

[17] Zhang Q., Guan J. Single-atom catalysts for electrocatalytic applications[J]. Adv. Funct. Mater., 2020, 30:2000768.

[18] Xia B., Zhang Y., Ran J., et al. Single-atom photocatalysts for emerging reactions[J]. ACS Cent. Sci., 2021, 7:39-54.

[19] Wang Y., Mao J., Meng X., et al. Catalysis with two-dimensional materials confining single atoms[J]. Chem. Rev., 2019,119:1806-1854.

[10] Zhang B., Fan T., Xie N., et al. Versatile applications of metal single-atom @ 2D material nano-platforms[J]. Adv. Sci., 2019, 6:1901787.

[21] Zhang W., eng Q., Shi L., et al. Merging single-atom-dispersed iron and graphitic carbon nitride to a joint electronic system for high-efficiency photocatalytic hydrogen evolution[J]. Small, 2019, 15:1905166.

[22] Jin X., Wang R., Zhang L., et al. Electron configuration modulation of Ni single-atoms for remarkably elevated photocatalytic hydrogen evolution[J]. Angew. Chem. Int. Ed., 2020, 59:6827-6831.

[23] Zhou P., Lv F., Li N., et al. Strengthening reactive metal-support interaction to stabilize high-density Pt single atoms on electron-deficient g-C₃N₄ for boosting photocatalytic H₂ production[J]. Nano Energy, 2019, 56:127-137.

[24] Zhang Q., Guan J. Recent progress in single-atom catalysts for photocatalytic water splitting[J]. Sol. RRL, 2020,4:2000283.

[25] Jiang H. X., Zhang S. L., Liu Y.H., et al. Silver single atom in carbon nitride catalyst for highly efficient photocatalytic hydrogen evolution angew[J]. Chem. Int. Ed., 2020, 59: 23112-23116.

[26] Shi R., Tian C., Zhu X., et al. Achieving an exceptionally high loading of isolated cobalt single atoms on porous carbon matrix for efficient visible-light-driven photocatalytic hydrogen production[J]. Chem. Sci., 2019,10: 2585-2591.

[27] Hu Y., Qu Y., Zhou Y., et al. Single Pt atom-anchored C₃N₄: A bridging Pt-N bond

boosted electron transfer for highly efficient photocatalytic H_2 generation Chem[J]. Eng. J. ,2021,412: 128749.

[28] Jia L., Cheng X., Wang X., et al. Large-scale preparation of g-C_3N_4 porous nanotubes with enhanced ind[J]. Eng. Chem. Res., 2020, 59:1065-1072.

[29] Xu J., Zhang S., Liu X., et al. Rh/polymeric carbon nitride porous tubular catalyst: visible light enhanced chlorophenol hydrodechlorination in base-free aqueous medium catalysis[J]. Sci. Technol., 2019,9: 6938-6945.

[30] Dong H., Guo X., Yang C., et al. Synthesis of g-C_3N_4 by different precursors under burning explosion effect and its photocatalytic degradation for tylosin Appl[J]. Catal. B-Environ., 2018,230: 65-76.

[31] Zhang G., Zhang J., Zhang M., et al. Polycondensation of thiourea into carbon nitride semiconductors as visible light photocatalysts[J]. J. Mater. Chem., 2012, 22:8083-8091.

[32] Chen Z., Pronkin S., Fellinger P. T., et al. Merging single atom-dispersed silver and carbon nitride to a joint electronic system via copolymerization with silver tricyano methanide [J]. ACS Nano, 2016,10: 3166-3175.

[33] Chen W. P., Li K., Yu X. Y., et al. Cobalt-doped graphitic carbon nitride photocatalysts with high activity for hydrogen evolution[J]. Appl. Surf. Sci. ,2017, 392:608-615.

[34] Gao J., Wang Y., Zhou S., et al. A facile one-step synthesis of Fe-doped g-C_3N_4 nanosheets and their improved visible light photocatalytic performances[J]. ChemCatChem., 2017,9: 1708-1715.

[35] Yu J., Chen G., Sunarso J., et al. Cobalt oxide and cobalt-graphitic carbon core-shell based catalysts with remarkably high oxygen reduction reaction activity[J]. Adv. Sci., 2016, 3:1600060.

[36] Wang P., Ren Y., Wang R. P., et al. Atomically dispersed cobalt catalyst anchored on nitrogen doped carbon nanosheets for lithium-oxygen batteries[J]. Nat. Comm. ,2020, 11:1576.

[37] Yang Y., Zeng G., Huang D., et al. In situ grown single-atom cobalt on polymeric carbon nitride with bidentate ligand for efficient photocatalytic degradation of refractory antibiotics[J]. Small, 2020,16: 2001634.

[38] Liu W., Hu W., Yang L., et al. Single cobalt atom anchored on carbon nitride with well-defined active sites for photo-enzyme catalysis[J]. Nano Energy, 2020, 73:104750.

[39] An Z., Gao J., Wang L., et al. Novel microreactors of polyacrylamide (PAM)CdS microgels for admirable photocatalytic H_2 production under visible light[J]. Int. J. Hydrog. Energy, 2019,44: 1514-1524.

[40] Cheng L., Xie S., Zou Y., et al. Noble-metal-free Fe_2P-Co_2P co-catalyst boosting visible-light-driven photocatalytic hydrogen production over graphitic carbon nitride: The synergistic effects between the metal phosphides[J]. Int. J. Hydrog. Energy, 2019,44: 4133-4142.

[41] Han M., Wang H., Zhao S.,et al. One-step synthesis of CoO/g-C_3N_4 composites by thermal decomposition for overall water splitting without sacrificial reagents. Inorg[J]. Chem. Front., 2017, 4:1691-1696.

[42] Liu X., Zhang Q., Liang L., et al. In-situ growing of CoO nanoparticles on g-C_3N_4

composites with highly improved photocatalytic activity for hydrogen evolution[J]. R. Soc. Open Sci., 2019, 6: 190433.

[43] Cao J. Y., Liu W., Cao L. L., et al. Atomic-level insight into optimizing hydrogen evolution pathway over a Co^1-N_4 single-site photocatalyst[J]. Angew. Chemie-Int. Ed., 2017, 56:12191-12196.

Chapter 9

B and P doped g-C₃N₄

9.1 Background

To obtain the g-C₃N₄ with excellent photocatalytic water splitting, non-metal doping into matrix of g-C₃N₄ to alter electronic structure has been proved to be an effective way to improve optical properties and charge separation rate to boost the H_2 or O_2 generation from water splitting. Lin et al. reported that B-doped could promote the red-shift of the optical absorption edge to upgrade photocatalytic water splitting performance by theoretical research[1]. Huang et al. adjusted the energy band structure of g-C₃N₄ via doping P to improve the O_2 production from photocatalytic water splitting [2]. Zhao et al. prepared the boron-doped carbon nitride with Z-scheme heterostructures for photocatalytic overall water splitting[3]. However, the insufficient O_2 production capacity is an impediment to upgrade the solar energy utilization efficiency of photocatalytic water splitting, and the addition of hole sacrificial agents (such as methanol, triethanolamine, sodium sulfite, etc.) is used to improve the driving force for water splitting, which is unfavourable for economy and environment because of the excessive consumption of expensive and valuable sacrificial agents. As low concentration of bisphenol A (BPA), one of endocrine-disrupting chemicals (EDCs), in the waterbody from manufacturing industry is identified as the micro-pollutant to cause respiratory diseases, and the low cost strategy is urgently needed to eliminate these EDCs in the water[4]. Therefore, the synergy between photocatalytic water splitting and wastewater treatment instead of sacrificial agent consumption is a promising way to improve the solar energy utilization efficiency of photocatalytic water splitting by carbon nitride.

In this book, B and P doped g-C₃N₄ photocatalysts were prepared by a simple and low-cost method to enhance the photocatalytic activity of water splitting under visible light. The as-prepared samples were characterized by state of the art tools to obtain the morphology, optical and photo-electrochemical properties. The photocatalytic water splitting activity and BPA degradation performance were carried out under the irradiation of simulated sunlight. The possible photocatalytic mechanisms were also investigated.

9.2   Facile synthesis of B and P doped g-C₃N₄

Melamine, sodium borohydride, sodium dihydrogen phosphate, bisphenol A and triethanolamine (TEOA) were purchased from Aladdin Industrial Corporation (Shanghai, China). All chemicals were used without further purification.

Bulk two dimensional g-C₃N₄ was synthesized according to previous method with slight modifications in the reported work[5]. In typical procedure, melamine (10g, 79.29 mmol) was put into covered crucible and subsequently calcined at 550 °C for 3h with heating rate of 10 °C /min in the muffle furnace to obtained bulk g-C₃N₄. After the material cooled to room temperature, the bulk g-C₃N₄ was ground into powder form with mortar and placed in the porcelain boat to be treated in a tube furnace. The g-C₃N₄ sheets were obtained by calcination under N₂ atmosphere at 600 °C for 2 h with the ramping rate of 2 °C/min.

The boron and phosphorus doped g-C₃N₄ was synthesized by using sodium borohydride and sodium dihydrogen phosphate as the B and P precursor. Specifically, 100mg g-C₃N₄ sheets and a certain amount of sodium borohydride or sodium dihydrogen phosphate were mixed together and ground by a mortar. After that, the mixture was calcined at 400 °C for 2 h with the ramping rate of 2 °C/min under N₂ atmosphere in a tube furnace. Then, the obtained solid was washed by water for several times and dried in the oven. After cooling to room temperature, these products were designated as CN-B, CN-P and CN-BP and kept in air tight glass vial for further characterization and photocatalysis. The mass fraction of B or P was 1% in the CN-B and CN-P, while 0.5% B and 0.5% P were co-doped in the g-C₃N₄ sheets to form CN-BP.

9.3   Characterization of B and P doped g-C₃N₄

9.3.1   XRD, SEM and TEM of B and P doped g-C₃N₄

The process for boron and phosphorus co-doped g-C₃N₄ is schematically shown in Figure 9.1. The crystalline structures of as-prepared materials were carried out by X-ray diffraction (XRD) patterns, as shown in Figure 9.2 (b). Generally, it is assumed that g-C$_3$N$_4$ is constructed by the tri-s-triazine building blocks. The intense peaks at 27.9° and 13.0° indexed for graphitic materials at (002) facet and (100) facet are the characteristic peaks for interlayer stacking aromatic systems of carbon nitride. After doping boron or phosphorus, the diffraction peak at 27.9° is a little shifted toward the lower 2 theta value as shown in Figure 9.2 (a), suggesting that boron or phosphorus have been successfully doped in the triazine ring of g-C$_3$N$_4$. In contrast, when B and P are co-doped in g-C$_3$N$_4$ sheets simultaneously, the same peak in CN-BP is slightly shifted toward a higher 2 theta value owing to the smaller

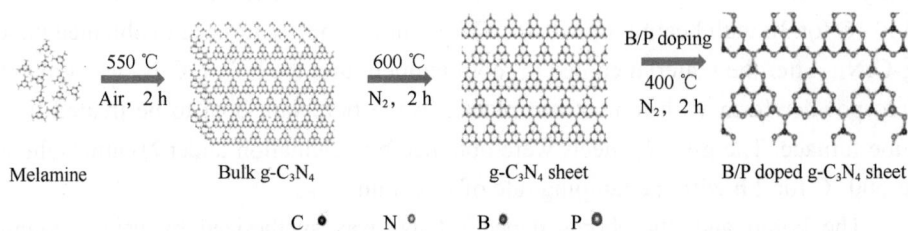

Figure 9.1   Graphic illustration of different process for the synthesis of B/P doped g-C$_3$N$_4$

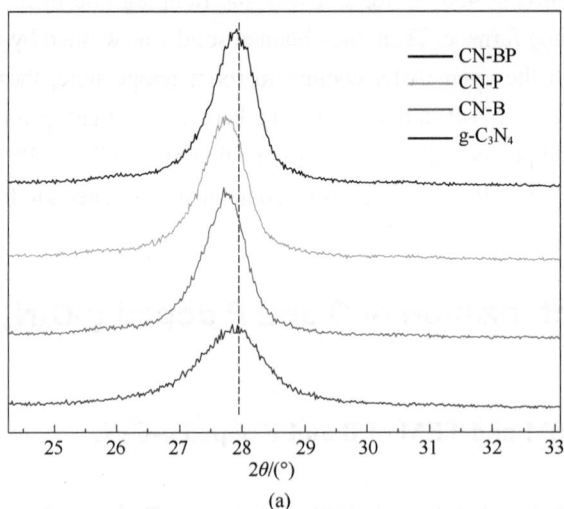

(a)

(b)

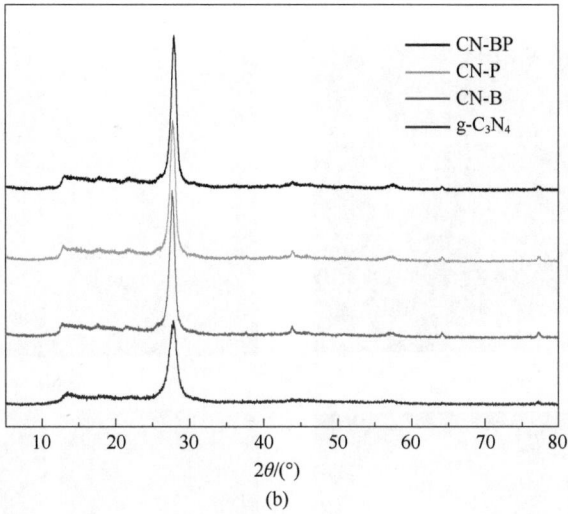

Figure 9.2　(a) XRD patterns of g-C₃N₄, CN-B, CN-P and CN-BP, (b) XRD patterns of g-C₃N₄, CN-B, CN-P and CN-BP

radius of B atom. Figure 9.3 shows the SEM and TEM images of g-C₃N₄ and CN-BP. In Figure 9.3(a) and Figure 9.3(c), the pure g-C₃N₄ possesses the layered structure of large particles. The CN-BP presents the similar layered structure of particles in Figure 9.3(b) and Figure 9.3(d). The element distribution reveals that the B and P elements are doped into the structure of g-C₃N₄ in Figure 9.3(e).

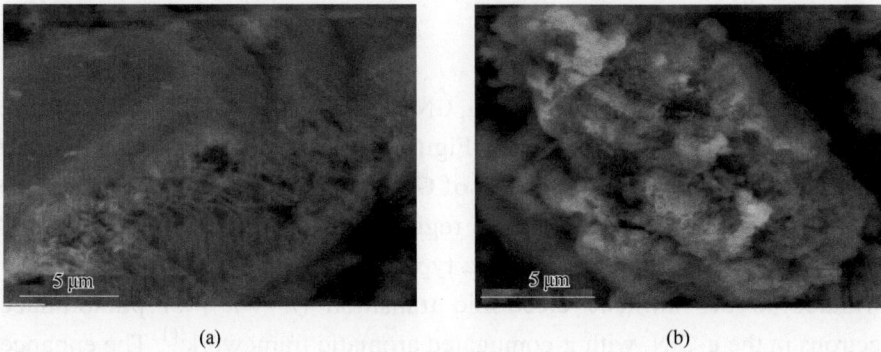

(a)

(b)

Figure 9.3

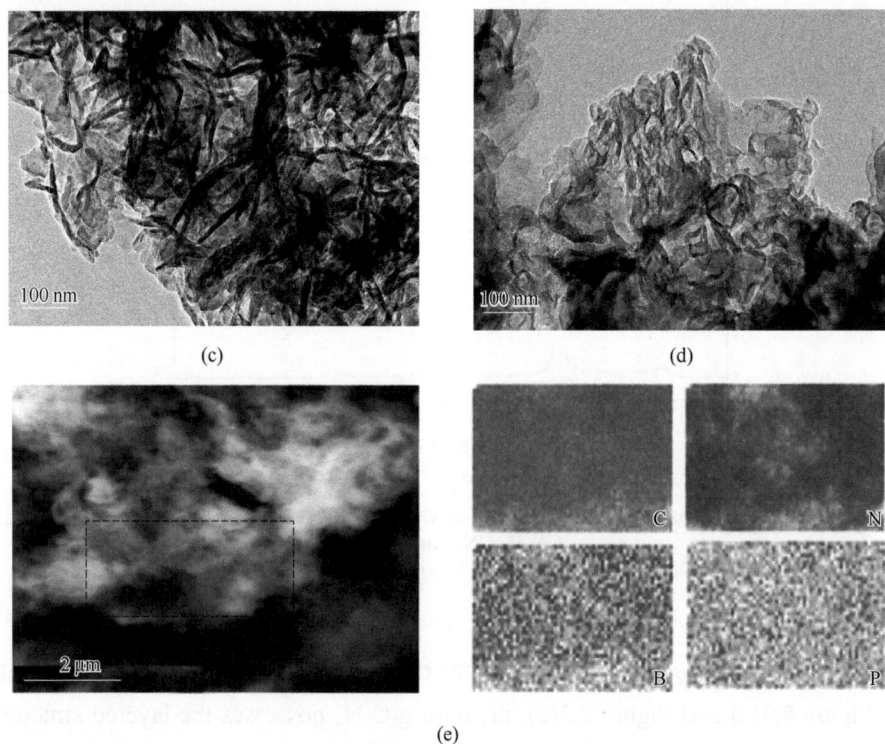

(c)

(d)

(e)

Figure 9.3   SEM and TEM images of g-C_3N_4 (a, c) and CN-BP (b,d) and the element mappings of CN-BP (e)

9.3.2   UV-vis absorption

The optical absorption of g-C_3N_4, CN-B, CN-P and CN-BP was investigated by UV-vis spectrum as displayed in Figure 9.4 (a). Compared to g-C_3N_4, after B or P doping , the light absorption of CN-B, CN-P and CN-BP was enhanced in both ultraviolet and visible-light regions. Fresh g-C_3N_4 shows negligible light absorbance above 500 nm and a typical absorption edge around 460 nm, attributed to the intrinsic electronic transition ($\pi \rightarrow \pi^*$) of photoinduced electrons in the g-C_3N_4 with π-conjugated aromatic framework[6]. The enhanced electronic transition ($\pi \rightarrow \pi^*$) suggests the better packing of the joint heptazine structure in the CN-B, CN-P and CN-BP. In addition, a significant change of absorption response at around 490 nm appears in the UV-vis spectrum of CN-B, CN-P and CN-BP, ascribed to the $n \rightarrow \pi^*$ electronic transition[7], indicating that the symmetric and planar heptazine structures of g-C₃N₄ have been destroyed by B and P doping. Figure 9.4 (b) (supporting information) presents the bandgaps of g-C₃N₄, CN-B, CN-P and CN-BP. After B and P doping, the bandgaps of samples decrease in comparison with bandgaps of

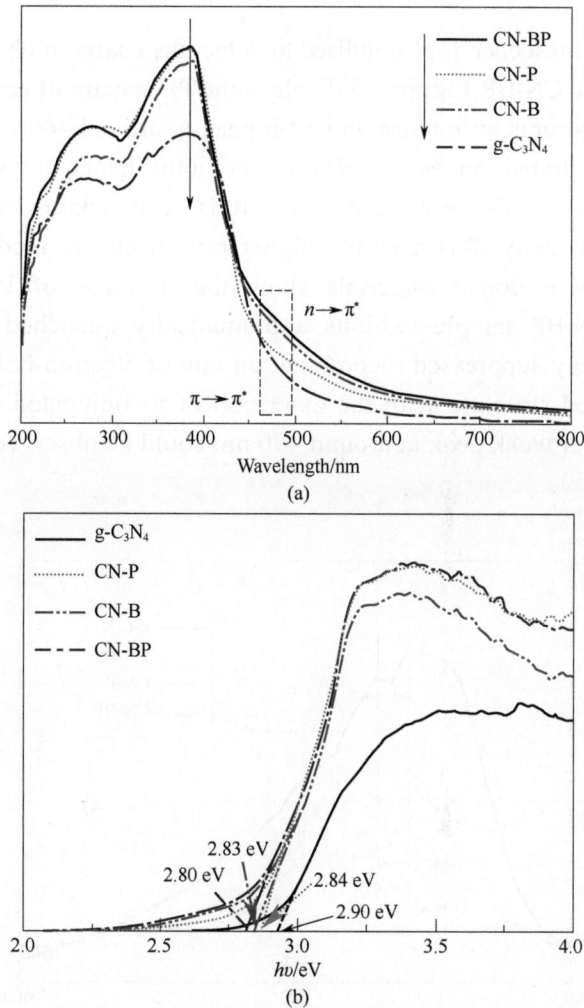

Figure 9.4   (a) UV-vis absorption spectra of the synthesized g-C₃N₄, CN-B, CN-P and CN-BP, (b) lots of $(\alpha h \upsilon)^2$ versus photon energy $(h\upsilon)$ of g-C₃N₄, CN-B, CN-P and CN-BP

147

pure g-C_3N_4. The reduced bandgaps of CN-B, CN-P and CN-BP are attributed to the improvement of charge mobility and extension of π-conjugated aromatic framework, contributing to the easier activation of intrinsic electronic transition ($\pi \rightarrow \pi^*$). The results demonstrate that n $\rightarrow \pi^*$ electronic transition of g-C_3N_4 could be awaked feasibly by B and P doping, and then, the photoresponse range is significantly expanded up to 600 nm.

9.3.3   PL spectra of B and P doped g-C_3N_4

Photoluminescence is also utilised to detect the charge mobility of g-C_3N_4, CN-B, CN-P and CN-BP. Figure 9.5 displays the PL spectra of prepared samples. Pure g-C_3N_4 presents an intense emission peak at around 460 nm, ascribed to the electronic transition ($\pi \rightarrow \pi^*$) in the joint heptazine structure with π-conjugated aromatic ring framework. It is well acknowledged that the stronger PL intensity illustrates the higher rate of charge recombination. In contrast, B or P doped materials show the decrease of PL quenching. Especially, CN-BP sample exhibits a dramatically quenched PL intensity, signifying highly suppressed recombination rate of electron-hole pairs owing to the improved structure with the extension of π-conjugated aromatic ring. Besides, another weak peak at around 490 nm could be observed in the B or P

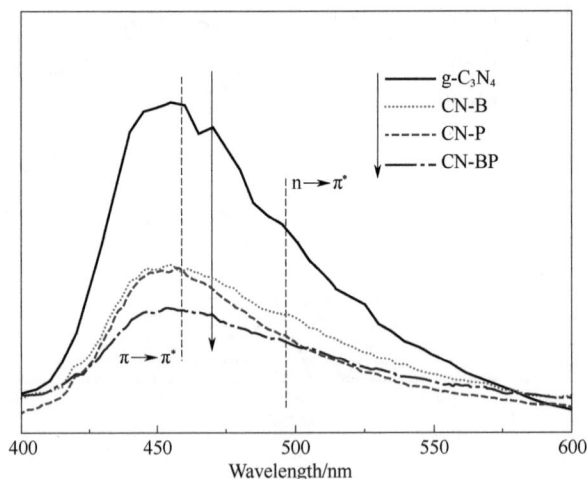

Figure 9.5   PL spectra ($\lambda_{ex} = 380$ nm) for the synthesized g-C_3N_4, CN-B, CN-P and CN-BP

doped materials, deconvoluted to the n → π* electronic transition, in agreement with the UV-vis absorbance results. Therefore, it can be deduced that the utilization of enhanced visible-light absorption could be improved originated from the n → π* electronic transition of g-C₃N₄ by B or P doping and the photocatalytic H_2 evolution could be reasonably enhanced.

9.3.4   XPS of B and P doped g-C₃N₄

XPS analysis are further conducted to detect detailed chemical compositions of g-C₃N₄, CN-B, CN-P and CN-BP, and the XPS results are presented in Figure 9.6. In Figure 9.6 (f) (supporting information), there are three obvious peaks around 285 eV, 399 eV and 530 eV of g-C₃N₄, CN-B, CN-P and CN-BP could be corresponding to the signals of C 1s, N 1s and O 1s, respectively. The XPS survey spectra of prepared samples with weak peaks around 191 eV and 133 eV indicate the presence of B and P. Table 9.1 shows atomic relative content (%) of g-C₃N₄, CN-B, CN-P and CN-BP from XPS characterization, and the XPS results reveal that B and P elements have been successfully doped into the joint heptazine structure of g-C₃N₄. Figure 9.6 (a) presents the XPS spectra of C 1s of g-C₃N₄, CN-B, CN-P and CN-BP. The C 1s spectra can be divided into three peaks in all the prepared samples and it is found that three peaks around 294.1 eV, 287.8 eV and 284.3 eV are corresponding to the signals of π-excitation, the sp^2 bonded C of N=C—N group in the tir-s-triazine rings and the surface carbon of C-C coordination, respectively[8]. The relative content of C-C in doped sample is remarkably lower than that of pure g-C₃N₄, while the relative content of N=C—N in doped sample is higher than pure g-C₃N₄, as shown in Table 9.2. The N 1s spectra of g-C₃N₄, CN-B, CN-P and CN-BP can be deconvoluted into four peaks at about 398.2 eV, 398.8 eV, 400.3 eV, and 404.2 eV, as displayed in Figure 9.6 (b), which are ascribed to the signals of nitrogen in the C=N—C group[9], bonded tertiary nitrogen in the N-C₃ group[10], nitrogen in the amino function group[11] and nitrogen with positive charge effects in heterocycle[12], respectively. It can be seen that in the Figure 9.6 (c) the binding energy of B 1s is located at 191.0 eV, slightly higher than the h-BN (190.0 eV) and lower than BCN(H) (192.1 eV), indicating the B atom is doped into CN by substituting C[13]. Besides, the binding energy of P 2p, shown in Figure 9.6 (d), is centered at 133.1 eV, which is typical for a P-N coordination. Therefore, it is assumed that B and P are successfully incorporated into the joint heptazine ring by substituting C atoms. It was reported that B or P was attached with three nitrogen

149

atoms in the ring based on the first-principle density functional theory (DFT)[14]. The relative contents of different nitrogen groups of the prepared samples are shown

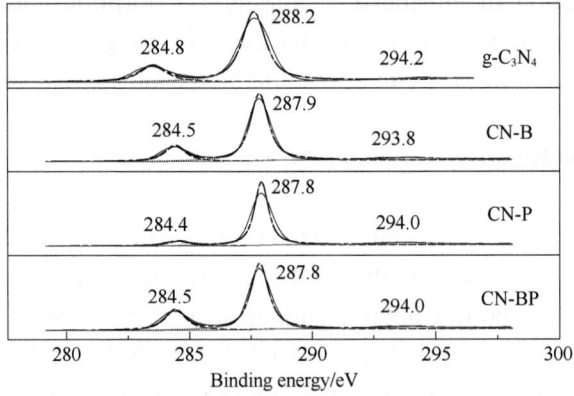

(a)

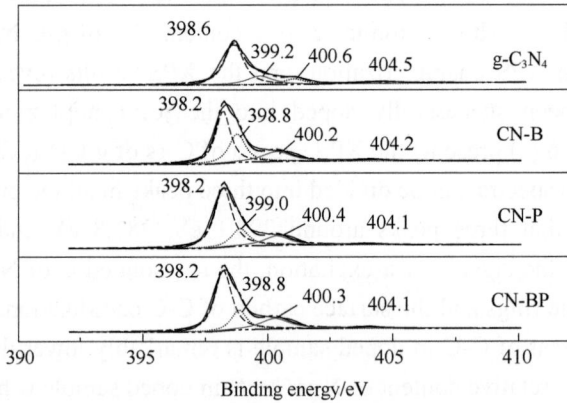

(b)

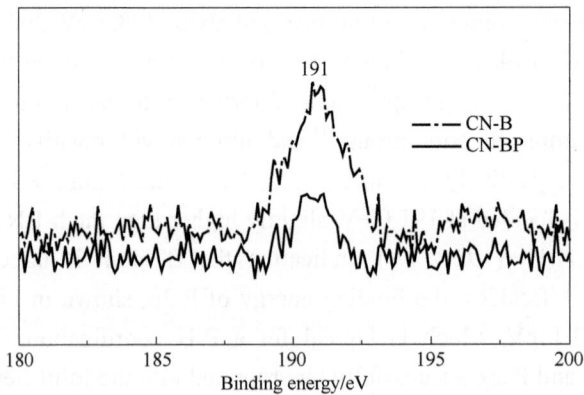

(c)

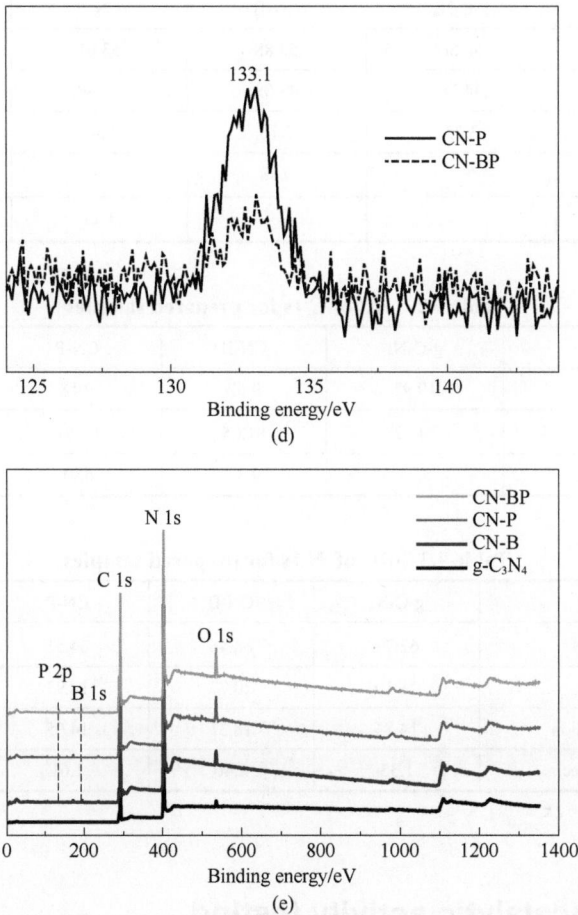

Figure 9.6   XPS profiles of (a) C 1s, (b) N 1s, (c) B 1s and (d) P 2p of the prepared g-C₃N₄, CN-B, CN-P and CN-BP, (e) XPS survey scan of g-C₃N₄, CN-B, CN-P and CN-BP

in Table 9.3. The heptazine-based structure of B and P doped g-C₃N₄ is consistent with the N 1s XPS spectrum, and the C=N—C/N-C₃ ratio determines the distortion and deformation degree of the joint heptazine ring, which will promote the n → π* electronic transition in g-C₃N₄[7]. According to the results in Table 9.3, the C=N—C/N-C₃ ratio of B and P doped g-C₃N₄ is significantly higher than that of pure g-C₃N₄, and the C=N—C/N-C₃ ratio of CN-BP is highest, indicating the best photocatalytic activities among the prepared samples.

Table 9.1   Atomic relative content (%) of prepared samples from XPS characterization

Sample	g-C$_3$N$_4$	CN-B	CN-P	CN-BP
N	52.66	52.88	53.01	52.23
C	44.28	43.08	42.40	42.98
O	3.05	3.16	3.42	3.43
B	—	0.88	—	0.64
P	—	—	1.17	0.72

Table 9.2   RC of C 1s for prepared samples

Sample	g-C$_3$N$_4$	CN-B	CN-P	CN-BP
C-C	19.93	9.62	9.18	14.42
N=C—N	74.32	81.05	83.91	79.01
π-excitation	5.75	9.33	6.91	6.57

Table 9.3   RC of N 1s for prepared samples

Sample	g-C$_3$N$_4$	CN-B	CN-P	CN-BP
N=C—N	62.73	68.46	64.51	61.62
N-(C)$_3$	21.10	20.77	18.93	18.04
Amino function	14.84	18.31	14.95	18.33
Charging effects	1.33	2.46	1.61	2.01

9.4   Photocatalytic activity testing

9.4.1   Hydrogen production efficiency

Photocatalytic H$_2$ production studies were measured by an on-line analysis system (CEL-SPH2N, AG, CEAULIGHT, Beijing) in the reaction vessel with top-irradiation. 25 mg of photocatalytic material was added into 50mL aqueous solution contained TEOA and water. 1.5 wt% of Pt nanoparticle was in-situ doped on the surface of the samples by photodeposition process from the precursor H$_2$PtCl$_6$·6H$_2$O. The reaction vessel temperature was kept at about 10 °C by a flow of cooling water. Before the irradiation experiments, the reaction system was evacuated for 30 min to remove air. The visible light source was the 300 W xenon lamp with a 420 nm filter. The amount of generated hydrogen was determined by gas chromatography (GC 7920, Beijing) by using nitrogen as the carrier gas. The photocatalytic O_2 hydrogen activities of the prepared samples were measured in $AgNO_3$ (10 mM) solution with the co-catalyst (0.3 wt%) Co. The temperature of reaction vessel was kept at about 10°C by a flow of cooling water. Before the irradiation experiments, the reaction system was evacuated for 30 min to remove air. The visible light source was the 300 W xenon lamp with a 420nm cut-off filter. The amount of generated hydrogen was determined by gas chromatography (GC 7920, Beijing) by using nitrogen as the carrier gas.

Apparent quantum efficiency (AQE) was measured under the same photocatalytic reaction conditions, except that a 420 nm band pass filter was placed between the lamp and the reaction vessel. The average intensity and irradiation area were measured. The AQE was calculated according to the following equation:

$$AQE\ (\%)= \frac{number\ of\ reacted\ electrons}{number\ of\ incident\ photons} *100$$

$$= \frac{number\ of\ evolved\ gas\ molecules * n}{number\ of\ incident\ photons} *100$$

where $n=2$ for H_2 production and 4 for O_2 production.

The photocatalytic performance of g-C₃N₄, CN-B, CN-P and CN-BP was evaluated by monitoring H_2 or O_2 evolution from water splitting under visible-light illumination ($\lambda \geqslant 420$ nm). Controlling experiments with absence of light illumination, no H_2 and O_2 evolution was detected. Figure 9.7 (a) and Figure 9.7 (b) display the temporal photocatalytic H_2 and O_2 evolution of g-C₃N₄, CN-B, CN-P and CN-BP where 0.3 wt% Pt and 0.3 wt% Co were loaded onto the surface of the prepared powders as co-catalyst for hydrogen and oxygen evolution, respectively. Oxygen production reaction is considered as the rate-limiting process for water splitting because it is thermodynamically and kinetically difficulty for involving participations of four holes and four protons. Obviously, B and P doping materials demonstrate much higher activities than fresh g-C₃N₄ for both H_2 and O_2 evolution. CN-BP exhibits the best activity and produces hydrogen and oxygen at a rate of 4.0 $mmol_{H_2}/(h \cdot g)$ and 0.3 $mmol_{O_2}/(h \cdot g)$, respectively. The apparent quantum efficiency (AQE) of hydrogen and oxygen evolution for CN-BP were estimated to be 0.35% and 0.20%.

As no hydrogen evolution was detected without sacrificial agent, the synergy between photocatalytic water splitting and BPA degradation is carried out to improve the solar energy utilization efficiency. The prepared samples presents outstanding photocatalytic activity for BPA degradation and hydrogen evolution, as shown in Figure 9.7 (c). Remarkably, the CN-BP could remove BPA completely and produce 30 $\mu mol/(h \cdot g)$ hydrogen. The enhanced photocatalytic activity could be assumed by the excellent optical properties and separation rate of photoexcited charge-hole carriers.

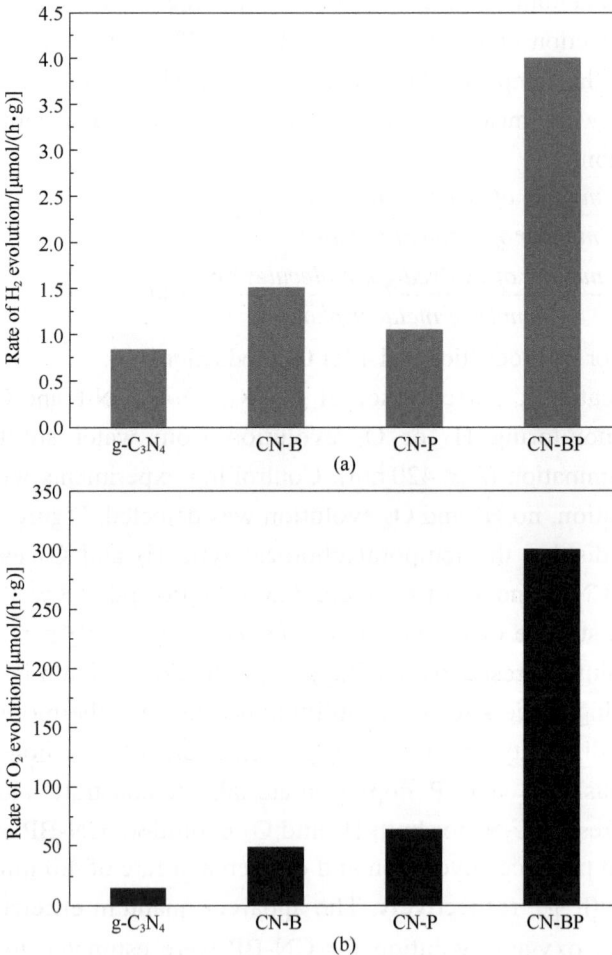

(a)

(b)

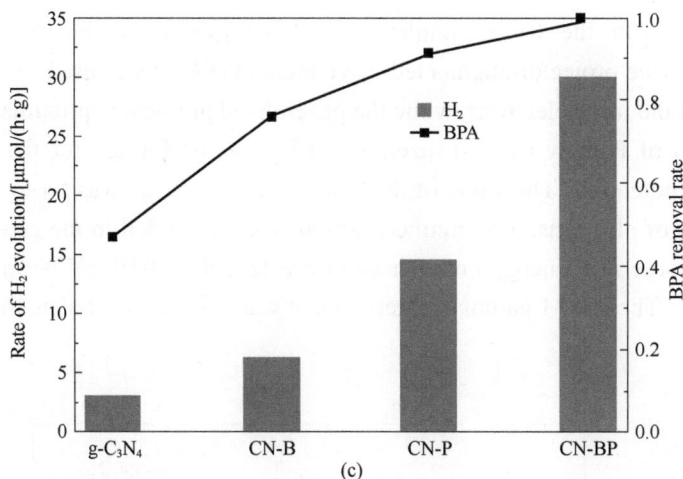

Figure 9.7   Photocatalytic activities of (a) hydrogen evolution, (b) oxygen evolution, (c) BPA degradation of the prepared g-C₃N₄, CN-B, CN-P and CN-BP

9.4.2   Charge separation and transfer performance

Electron paramagnetic resonance (EPR) spectra of g-C₃N₄ and CN-BP are evaluated under room temperature to further prove the n → π* electronic transition, as shown in Figure 9.7 (a). Compared to pure g-C₃N₄, EPR intensity of CN-BP is upgraded and the g-value is almost no change, indicating the existence of a lot of free charge carriers in the aromatic ring[15], and lots of free charge carriers could be generated from the lone pair of electrons excited through n → π* electronic transition. Therefore, B and P doped g-C₃N₄ is conducive to more extension of π-conjugated aromatic ring and greater mobility of the spins. The CN-BP sample also exhibit decreased electric resistance versus that of pure g-C₃N₄, as represented by the smaller radius of electrochemical impedance spectroscopy (EIS) in Figure 9.7 (b), suggesting the boosted charge mobility in CN-BP. Figure 9.8 (c) presents the Mott-Schottky (MS) plots of g-C₃N₄ and CN-BP. According to the Figure 9.8 (c), the CB potentials of g-C₃N₄ and CN-BP were obtained from the MS plots with flat-band potentials. The doping B and P elements can impact the bandgap structure of g-C₃N₄ by adjusting the electronic structure.

Geometry optimization and electronic structure calculations were performed using DFT with the Code simulation package (VASP version 5.4.4). In the calculations, the projector-augmented-wave method (PAW) was used for the pseudo potential of the inner electrons, while the generalized gradient approximation (GGA) in the form of Perdew-Burke-Ernzerhof (PBE) was performed for the exchange-correlation potentials. The value of the kinetic cut-off energy was set to 500 eV for the valence of electrons. The atomic positions are suspended in the configurations when the force and energy on each atom are less than 0.01 eV/Å and 10^{-5} eV, respectively. The 3×3×1 gamma-centered k-net was chosen to simulate the Brillouin

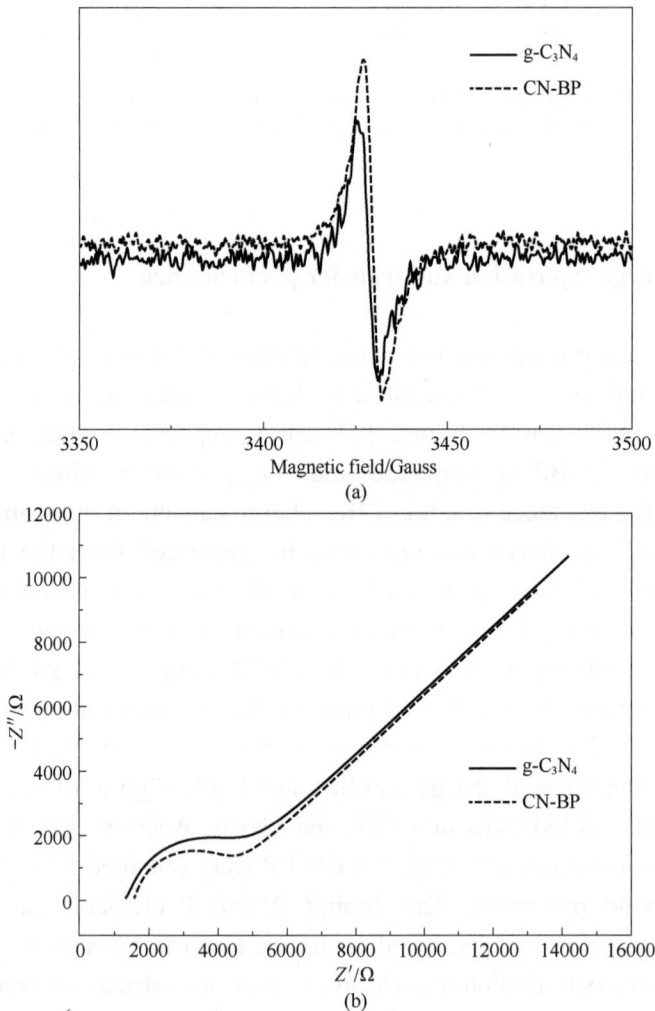

(a)

(b)

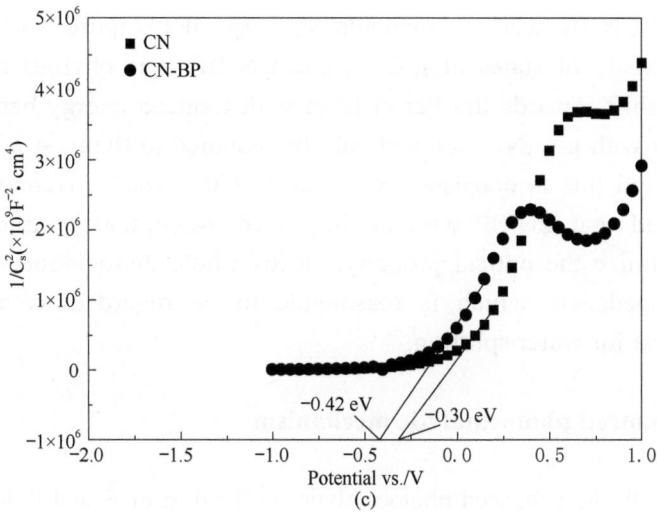

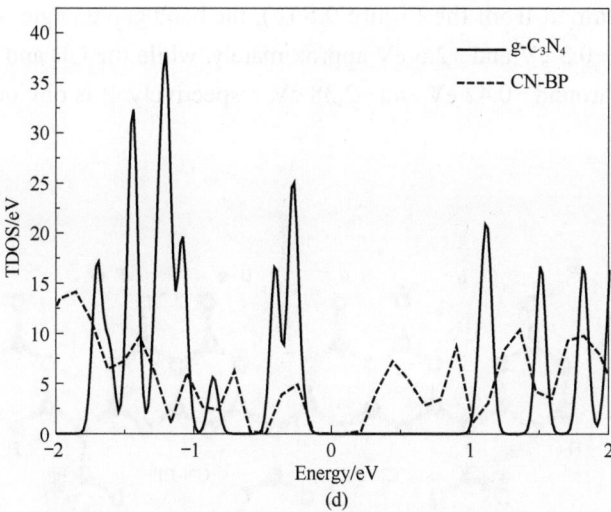

Figure 9.8   (a) EPR, (b) electrochemical impedance spectra (c) Mott-Schottky plots of the as-prepared g-C₃N₄ and CN-BP at -0.4 V versus Ag/AgCl, (d) Calculated total density of states for g-C₃N₄ and CN-BP

zone of the surface. The model of a two-dimensional graphitic carbon nitride was constructed with a 3×3 supercell. The B and P doped graphitic carbon nitride was simulated by substituting C atom in g-C₃N₄. The electrical band structures of g-C$_3$N$_4$ and CN-BP were also calculated using DFT. Figure 9.8 (d) presents the total density of states of g-C$_3$N$_4$ and CN-BP. It is obvious that CN-BP displays a shift towards the Fermi level with reduced energy band width in comparison with g-C$_3$N$_4$, which should be assumed to the n → π* electronic transition, and this is consistent with the UV-Vis results. Hereto, it can be demonstrated that CN-BP with an improved π-conjugated structure could overall optimize the optical property, electron-hole dissociation, and charge transfer impedance, which is reasonable to be regarded as an efficient photocatalyst for water splitting.

9.4.3 Enhanced photocatalytic mechanism

Above all, the enhanced photocatalytic mechanism of B and P doped carbon nitride for water splitting and BPA degradation is proposed in Figure 9.9. According to the bandgaps obtained from the Figure 9.4 (b) and the CB potentials obtained from the Figure 9.8 (c), the band gap energies of CB and VB of g-C$_3$N$_4$ are −0.3 eV and +2.6 eV approximately, while the CB and VB potentials of CN-BP are around −0.42 eV and +2.38 eV, respectively. It is obvious that the CB

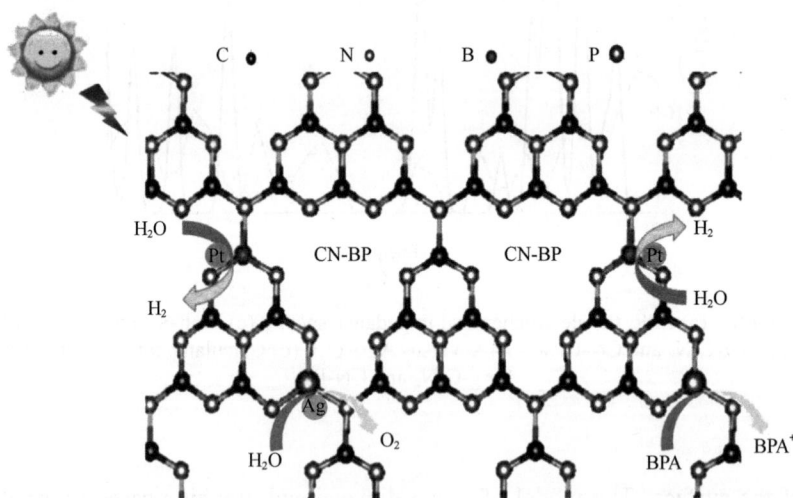

Figure 9.9   Photocatalytic mechanism of CN-BP for water splitting and BPA degradation

of CN-BP is more negative than that of g-C$_3$N$_4$. Owing to a sufficiently potential difference, the photogenerated electrons accumulated on the CB of CN-BP could be transferred to the Pt co-catalyst to generate H$_2$. The holes in VB of samples are interaction with the Ag co-catalyst to produce O$_2$. Furthermore, the BPA will be removal by the holes in VB of samples without the Ag co-catalyst. The incorporation of B and P atoms could adjust the electronic structure of g-C$_3$N$_4$ to enhance the separation efficiency of photogenerated charge carriers.

9.5 Conclusion

The fabrication of B and P doped carbon nitride as a highly efficient visible-light driven photocatalyst has been demonstrated by a facile, high-efficiency, and environmentally friendly two-step process. According to the detailed optical, and photoelectrochemical characterizations and DFT calculations, B and P doped g-C$_3$N$_4$ provide fast transfer of photoexcited electrons and an increased separation rate of e$^-$/h$^+$ pairs to boost photocatalytic water activity with extension of π-conjugated aromatic framework. The as-prepared CN-B, CN-P and CN-BP photocatalysts exhibit a higher photocatalytic H$_2$ and O$_2$ production than that of bulk g-C$_3$N$_4$. B and P doped carbon nitride present the enhanced synergetic activity between photocatalytic hydrogen evolution and BPA degradation, and the CN-BP photocatalyst could completely remove BPA to produce 30 μmol H$_2$ per hour. The result provides a new method for promoting the practical application of photocatalysts.

Reference

[1] Lin Y., Wang Q., Ma M., et al.Enhanced optical absorption and photocatalytic water splitting of g-C$_3$N$_4$/TiO$_2$ heterostructure through C&B codoping: A hybrid DFT study[J]. International Journal of Hydrogen Energy, 2021, 46 (14): 9417-9432.

[2] Huang M., Ai Z., Xu L., et al.Band structure-controlled P-C$_3$N$_4$ for photocatalytic water splitting via appropriately decreasing oxidation capacity[J]. Journal of Alloys and Compounds, 2022, 895: 162513.

[3] Zhao D., Wang Y., Dong C. L., et al.Boron-doped nitrogen-deficient carbon nitride-based Z-scheme heterostructures for photocatalytic overall water splitting[J]. Nature Energy, 2021, 6 (4): 388-397.

[4] Sahu R. S., Shih Y. H., Chen W. L.New insights of metal free 2D graphitic carbon nitride for photocatalytic degradation of bisphenol A[J]. Journal of Hazardous Materials, 2021, 402: 123509.

[5] Liu X., Mateen M., Cheng X., et al.Constructing atomic Co_1-N_4 sites in 2D polymeric carbon nitride for boosting photocatalytic hydrogen harvesting under visible light[J]. International Journal of Hydrogen Energy, 2022, 47 (25): 12592-12604.

[6] Su C., Zhou Y., Zhang L., et al.Enhanced n→π* electron transition of porous P-doped g-C_3N_4 nanosheets for improved photocatalytic H_2 evolution performance[J]. Ceramics International, 2020, 46 (6): 8444-8451.

[7] An S., Zhang G., Li K.,et al.Self-Supporting 3D Carbon nitride with tunable n → π* electronic transition for enhanced solar hydrogen production[J]. Advanced Materials, 2021, 33 (49): 2104361.

[8] Liu X., Zhang Q., Liang L., et al.In-situ growing of CoO nanoparticles on g-C_3N_4 composites with highly improved photocatalytic activity for hydrogen evolution[J]. Royal Society Open Science, 2019, 6 (7): 190433.

[9] Gao D., Xu Q., Zhang J., et al.Defect-related ferromagnetism in ultrathin metal-free g-C_3N_4 nanosheets[J]. Nanoscale, 2014, 6 (5): 2577-2581.

[10] Liu X., He L., Chen X., et al.Facile synthesis of CeO_2/g-C_3N_4 nanocomposites with significantly improved visible-light photocatalytic activity for hydrogen evolution[J]. International Journal of Hydrogen Energy, 2019, 44 (31): 16154-16163.

[11] Shen R., He K., Zhang A.,et al.In-situ construction of metallic Ni_3C@Ni core-shell cocatalysts over g-C_3N_4 nanosheets for shell-thickness-dependent photocatalytic H_2 production[J]. Applied Catalysis B: Environmental, 2021, 291:120104.

[12] Wang S., Li, C., Wang, T., et al., Controllable synthesis of nanotube-type graphitic C_3N_4 and their visible-light photocatalytic and fluorescent properties.Journal of Materials Chemistry A 2014, 2 (9): 2885-2890.

[13] Wang Y., Li H., Yao J., et al.Synthesis of boron doped polymeric carbon nitride solids and their use as metal-free catalysts for aliphatic C—H bond oxidation[J]. Chemical Science, 2011, 2 (3): 446-450.

[14] Liu Y., Zheng Y., Zhang W., et al.Template-free preparation of non-metal (B, P, S) doped g-C_3N_4 tubes with enhanced photocatalytic H_2O_2 generation[J]. Journal of Materials Science & Technology, 2021, 95: 127-135.

[15] Zhang G., Li G., Heil T., et al.Tailoring the grain boundary chemistry of polymeric carbon nitride for enhanced solar hydrogen production and $CO_{(2)}$ reduction[J]. Angewandte Chemie (International ed. in English), 2019, 58 (11): 3433-3437.

Chapter 10

Molten salt preparation of Fe@C$_3$N$_4$ nanosheets

10.1 Background

It is acknowledged that seawater possesses 97% of water resources on the earth. Therefore, the efficient solar-light-driven conversion of seawater into hydrogen energy is considered to be a more economical method to solve the global energy crisis and environmental problems. Subsequently, numerous semiconductor photocatalysts have been reported for seawater splitting and have made outstanding progress[1]. However, unsatisfactory factors such as ultraviolet-light-driven expenses and toxicity in seawater, affect the efficiency of photocatalytic hydrogen evolution and limit practical applications. Thus, visible-light-driven, low-cost, and stable semiconductors should be explored for photocatalytic seawater splitting.

According to the explored semiconductor photocatalysts, graphitic carbon nitride has been successfully utilized in seawater splitting because of the advantages of its excellent catalytic performance, suitable band gap, low cost, environmentally friendly nature, and high stability. Nevertheless, bulk g-C$_3$N$_4$ (BCN) displays poor photocatalytic seawater splitting activity owing to the poor utilization of visible light and severe recombination of photogenerated carriers with the presence of weak van der Waals forces in the adjacent layers with π-conjugated rings[2]. Several methods, such as metal doping, defect engineering, and heterojunction construction, have been unremittingly explored to enhance the photocatalytic seawater splitting activity of BCN[3].

To achieve excellent photocatalytic seawater splitting over BCN, high-temperature synthesis in salt melts and metal doping have been used to alter electronic structure to upgrade the photocatalytic efficiency of BCN photocatalyst. Liu et al. reported that carbon nitride by molten salt synthesis displayed good photocatalytic activity in simulated seawater[4]. Yang et al. prepared the triazine-/heptazine-based carbon nitride within a facile one-step molten salt route and achieved drastically enhanced photocatalytic hydrogen production performance[5]. It is reported that a carbon nitride-supported Ni-Fe oxide nanocluster is an efficient and durable anode for alkaline seawater oxidation[6]. However, the insufficient O_2 evolution performance impedes enhancing the energy conversion efficiency of photocatalytic seawater splitting, and the high activity performance is obtained in the presence of sacrificial electronic donors or hole' scavengers. However, these sacrificial agents may be quite expensive and would be consumed during several running recycles. The low concentration of glucosamine hydrochloride in the seawater body could be instead of sacrificial agent consumption for photocatalytic hydrogen evolution from seawater splitting.

In this book, Fe-doped MCN photocatalysts were prepared using a low-cost and simple method to enhance the photocatalytic activity of simulated seawater decomposition under visible light. The prepared samples were characterized using state-of-the-art tools to obtain morphological, optical, and electrochemical properties. The photocatalytic water decomposition activity of glucosamine hydrochloride as a sacrificial agent was only carried out under simulated sunlight irradiation. In addition, possible photocatalytic mechanisms were investigated.

10.2 Preparation of Fe@C₃N₄

Melamine, potassium chloride, lithium chloride, ferric nitrate, sodium chloride, and triethanolamine (TEOA) were purchased from Aladdin Industrial Corporation (Shanghai, China). All chemicals were used without further purification.

In a typical procedure, 3.00 g melamine was put into a covered crucible and subsequently calcined at 550 °C for 4 h with a heating rate of 10 °C /min in the muffle furnace to obtain bulk g-C_3N_4 was denoted as BCN. Firstly, 3.00 g of melamine was mixed with molten salts (1.83 g KCl and 1.51 g LiCl) to obtain modified molten-salt g-C_3N_4 was denoted as MCN. The mixture is then calcined under the same conditions. After the material cooled to room temperature, the bulk MCN was milled to powder form with a mortar and placed in the porcelain boat to be treated in a tube furnace.

The MCN-Fe samples were prepared as follows: Specifically, 100 mg of MCN tablets were mixed with different volumes of 0.5% Fe $(NO_3)_3$ solution (0.6 mL, 1 mL, 1.6 mL, 2 mL, 3 mL). The homogeneous mixture was then heated to 100 °C in a water bath to remove the water. After that, the dry mixture was ground and fired for 3 hours at 400°C in an N_2 atmosphere in a tube furnace at a rate of 2 °C per minute. After cooling down to room temperature, these products were stored in tightly sealed glass jars for further characterization and photocatalysis. Using this method, MCN-Fe samples were synthesized at different concentrations (0.3 g/mL, 0.5 g/mL, 0.8 g/mL, 1.0 g/mL, and 1.5 g/mL), which for simplicity can be denoted as 0.3MCN-Fe, 0.5MCN-Fe, 0.8MCN-Fe, 1.0MCN-Fe and 1.5MCN-Fe.

10.3   Characterization of Fe@C₃N₄

10.3.1   XRD, SEM and TEM of Fe@C₃N₄

MCN and Fe loading in MCN powder were prepared by the molten salt method, as shown in Figure 10.1 (a). The crystal structure of the produced materials was analyzed using XRD, as shown in Figure 10.2 (a). It is generally believed that BCN consists of tri-s-triazine building blocks. Three XRD peaks were present at 12.90°, 24.90° and 27.62°. This confirms that the synthesized sample was graphitized carbon nitride with a hexagonal structure, where the diffraction planes observed are (001), (101) and (002). In addition, (101) and (002) diffraction peaks slightly shift to the left side as Fe concentration increases, as shown in Figure 10.2 (b). The intensity of the above two peaks in the XRD pattern was slightly weakened by the addition of Fe ions, indicating that the Fe species were tightly bound to the MCN and therefore have an inhibitory effect on the structure. The diffraction peaks of the above photocatalysts showed that all the samples had good fusibility after the introduction of Fe ions, and the crystalline phase of MCN was not disrupted. Figure 10.3 shows the SEM and TEM images of MCN and the TEM plots of MCN-Fe. In Figure 10.3 (a) and Figure 10.3 (c), pure g-C₃N₄ (BCN) had a lamellar structure with large particles. MCN shows a more dispersed laminar structure in Figure 10.3 (b) and Figure 10.3 (d). The doping of Fe elements into the structure of MCN is shown in Figure 10.3 (e).

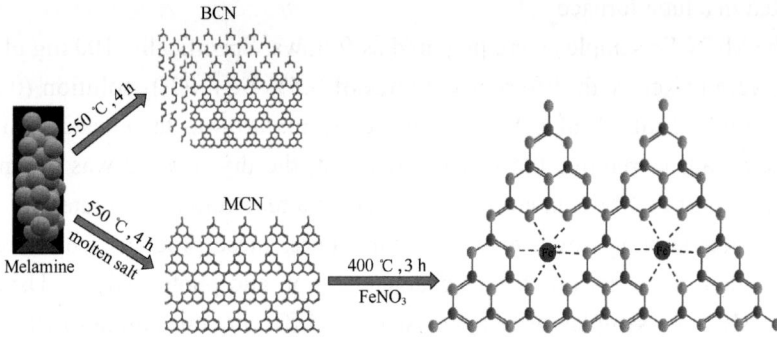

Figure 10.1　Graphicillustration of different carbon nitride preparation and iron loading processes

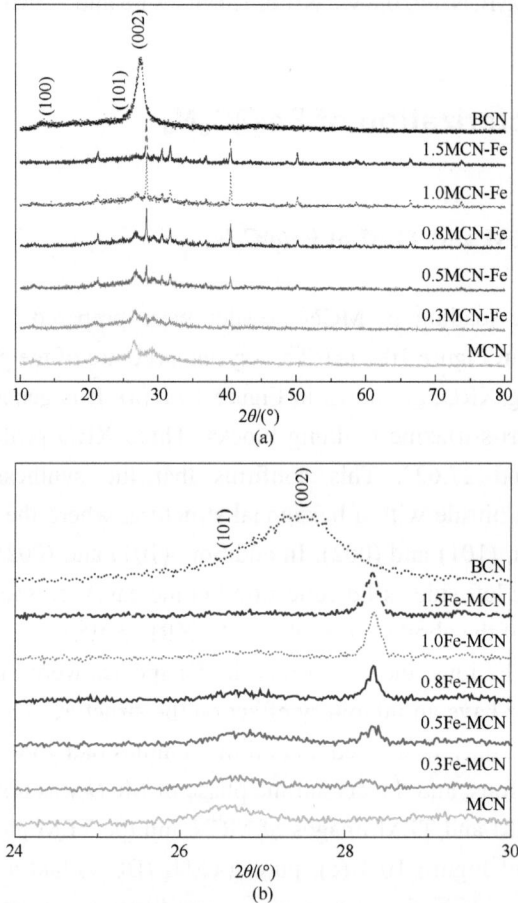

Figure 10.2　(a) XRD patterns of BCN, MCN, and patterns of Fe-doped MCN nanosheets with different Fe concentrations, (b) XRD patterns of BCN, MCN, and patterns of Fe-doped MCN nanosheets with different Fe concentrations

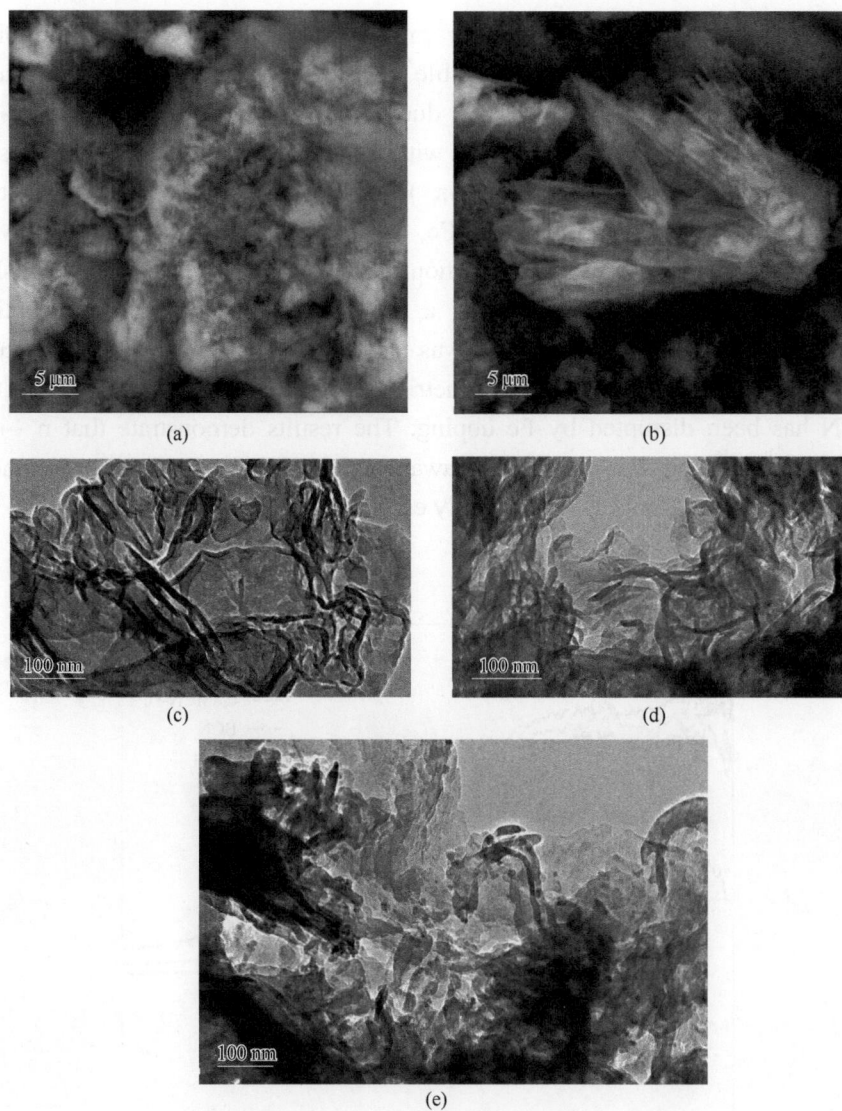

Figure 10.3   SEM and TEM images of BCN (a, c) and MCN (b, d) and the TEM image of Fe-doped MCN (e)

10.3.2   UV-vis spectroscopy of Fe@C$_3$N$_4$

The light absorption of BCN, MCN, 0.3MCN-Fe, 0.5MCN-Fe, and 0.8MCN-Fe was analyzed by UV-vis spectroscopy, as shown in Figure 10.4. Enhanced light absorption in the UV and visible regions of MCN, 0.3MCN-Fe,

0.5MCN-Fe, and 0.8MCN-Fe after molten salt method and iron doping compared to BCN. BCN had negligible light absorption above 500 nm and a typical absorption edge near 460 nm, due to inherent electron hopping ($\pi \rightarrow \pi^*$) of the light-induced electrons in BCN with π-conjugated aromatic backbones[7]. Enhanced electronic transitions ($\pi \rightarrow \pi^*$) indicate better stacking of absorption structures among MCN, 0.3MCN-Fe, 0.5MCN-Fe and 0.8MCN-Fe, with 0.5MCN-Fe being the best. In addition, the UV-Vis spectra of 0.3MCN-Fe, 0.5MCN-Fe, and 0.8MCN-Fe show a significant change in the absorption response around 490 nm, which was attributed to the n $\rightarrow \pi^*$ electron transitions[2] suggesting that the symmetric and planar heptazine structure of the MCN has been disrupted by Fe doping. The results demonstrate that n $\rightarrow \pi^*$ electronic transition of MCN could be awakened feasibly by Fe doping, and then, the photo response range is significantly expanded up to 600 nm.

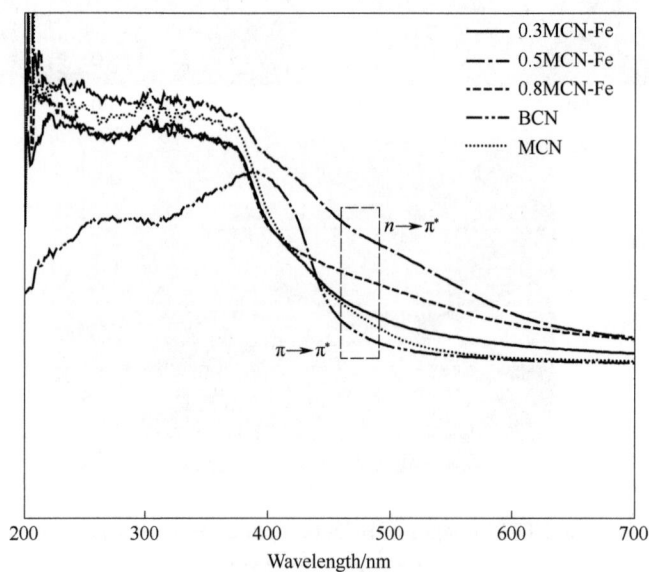

Figure 10.4   UV-vis absorption spectra of the synthesized BCN, MCN, 0.3MCN-Fe, 0.5MCN-Fe, and 0.8MCN-Fe.

10.3.3   PL spectroscopy of Fe@C₃N₄

Photoluminescence was also used to detect the charge mobility of BCN,

MCN, 0.3MCN-Fe, 0.5MCN-Fe, and 0.8MCN-Fe. The photoluminescence (PL) spectra of the prepared samples are shown in Figure 10.5. Pure BCN exhibited a strong emission peak near 460 nm attributed to an electron jump ($\pi \rightarrow \pi^*$) in the structure of the bound heptazine with a π-conjugated aromatic ring framework. On the contrary, the PL quenching rate was reduced for the molten salt method and iron-doped materials. In particular, the 0.5MCN-Fe sample exhibited a dramatically quenched PL intensity, indicating that the rate of electron-hole pair complexation was highly suppressed with the extension of the π-conjugated aromatic ring. In addition, another weak peak around 490 nm could be observed in the Fe-doped material, which de-convolves with the n $\rightarrow$ π^* electron leaps[8], in agreement with the UV-vis absorbance results. Therefore, it could be inferred that the utilization of enhanced visible light absorption could be improved by electron leaps of Fe-doped MCN and the photocatalytic H_2 evolution could be reasonably enhanced.

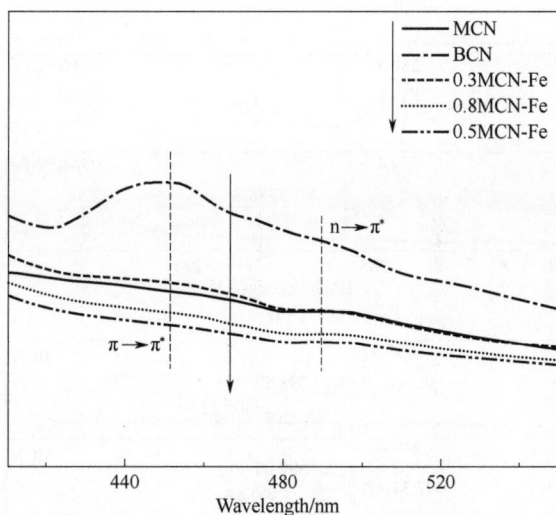

Figure 10.5   Photoluminescence (PL) spectra (λ_{ex} = 300 nm) for the synthesized BCM, MCN, 0.3MCN-Fe,0.5MCN-Fe, and 0.8MCN-Fe.

10.3.4   XPS of Fe@C₃N₄

Further XPS analysis was performed to detect the detailed chemical compositions of BCN, MCN, 0.5MCN-Fe, and 1.5MCN-Fe. The XPS results are shown in Figure 10.6. In Figure 10.6 (d) (supplementary information) the BCN shows three distinct peaks at 285 eV, 399 eV, and 530 eV, corresponding to C 1s, N 1s, and O 1s signals. Table 10.1 shows the atomic relative content (%) of BCN, MCN, 0.5MCN-Fe, and 1.5MCN-Fe from XPS characterization, and the XPS results reveal that Fe elements have been successfully doped into the joint heptazine structure of MCN. The C 1s spectra of all obtained samples can be divided into three peaks and it was found that the three peaks at 294.1 eV, 287.8 eV, and 284.3 eV correspond to the π-excitation signals, the sp2-bonded C in the N═C—N group of the tir-s-triazine ring and the surface carbon in the C—C coordination[9].

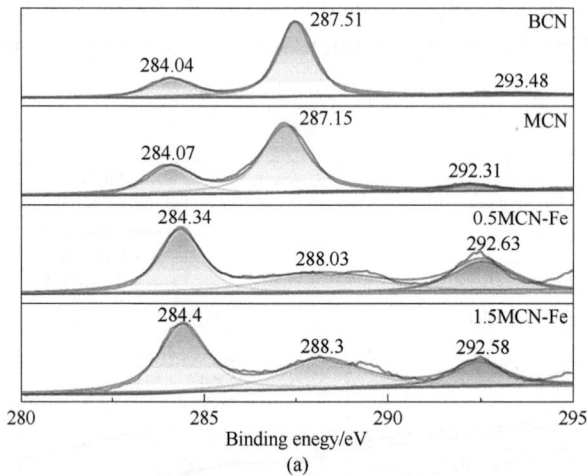

(a)

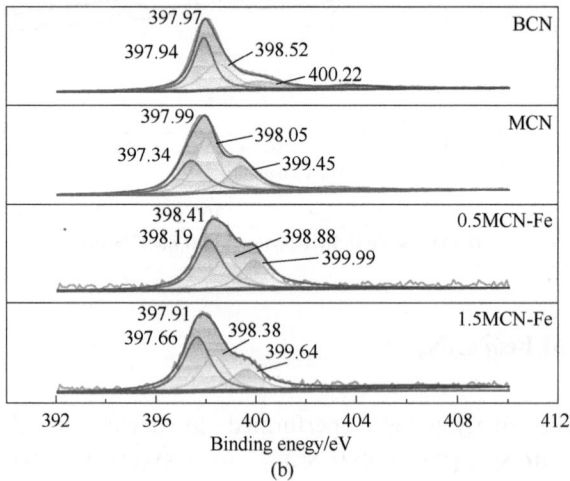

(b)

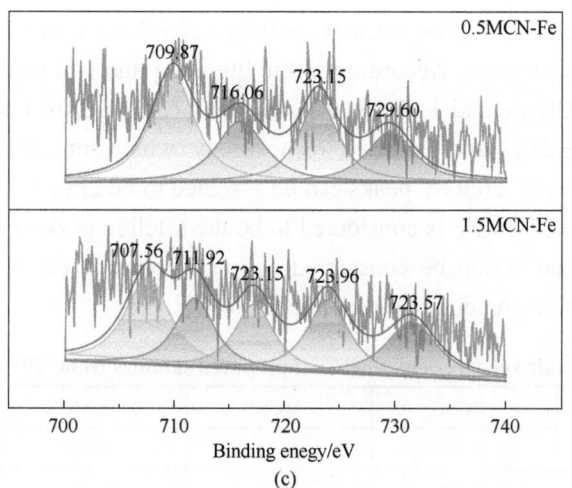

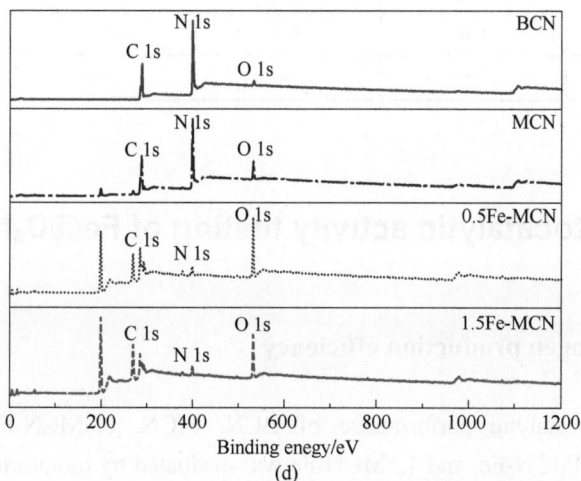

Figure 10.6   XPS profiles of (a) C 1s, (b) N 1s, (c)Fe 2p of the prepared BCN, MCN,0.5MCN-Fe and 1.5MCN-Fe, (d) XPS survey scan of BCN, MCN, 0.5Fe-MCN, and 1.5Fe-MCN

The peak of sp2 -bonded C in the N=C—N moiety of the tir-s-triazine ring after the addition of Fe is significantly shifted to the right by 0.49 eV shown in Figure 10.6 (a). The N 1s spectra of BCN, MCN, 0.5MCN-Fe, and 1.5MCN-Fe can be deconvoluted into four peaks at 398.2 eV, 398.8 eV, 400.2 eV, and 401.3eV as shown in Figure 10.6 (b), which are attributed to the nitrogen in the C=N—C moiety, the tertiary nitrogen bonded in the N-C₃ group the nitrogen in the amino-functional group and the nitrogen in the heterocyclic with a positive charge effect of

169

nitrogen[10]. The Fe 2p XPS spectra of the 0.5MCN-Fe and 1.5MCN-Fe samples are shown in Figure 10.6 (c). According to the literature, the two core-level signals of Fe in the 0.5MCN-Fe and 1.5MCN-Fe samples located at 709.3 and 723.2 eV are attributed to Fe $2P_{3/2}$ and Fe $2P_{1/2}$, respectively, which indicate the presence of Fe^{2+}. The 713.2 and 726.9 eV peaks can be assigned to Fe $2P_{3/2}$ and Fe $2P_{1/2}$ of Fe^{3+}, and the signal at 717.6 eV is considered to be the satellite peak of Fe^{2+} and Fe^{3+}. In this book pyridine N can be considered as metal-bound N(Fe-Nx), indicating the formation of a heterojunction between Fe and MCN.

Table 10.1　Atomic relative content (%) of prepared samples from XPS characterization

Sample	BCN	MCN	0.5Fe-MCN	1.5Fe-MCN
N	51.39	45.54	10.81	13.02
C	45.53	46.16	53.12	55.38
O	3.08	8.3	35.38	30.93
Fe	—	—	0.69	0.68

10.4　Photocatalytic activity testing of Fe@C₃N₄

10.4.1　Hydrogen production efficiency

The photocatalytic performance of BCN, MCN, 0.3MCN-Fe, 0.5MCN-Fe, 0.8MCN-Fe, 1.0MCN-Fe, and 1.5MCN-Fe was evaluated by monitoring the escape of H_2 or O_2 from water decomposition under visible light irradiation ($\lambda \geqslant 420$ nm). Figure 10.7 (a) and Figure 10.7 (b) display the temporal photocatalytic O_2 and H_2 evolution of BCN, MCN, 0.3MCN-Fe, 0.5MCN-Fe, 0.8MCN-Fe, 1.0MCN-Fe and 1.5MCN-Fe where 0.3 wt% of Pt was loaded onto the surface of the prepared powder as a hydrogen precipitation co-catalyst. It is clear that the molten salt method and iron-doped materials exhibit much higher activity than fresh BCN during the evolution of O_2 and 0.5MCN-Fe showed the best activity, producing oxygen at a rate of 123.58 μmol/(h·g) O_2. The apparent quantum efficiency (AQE) of hydrogen and oxygen evolution for 0.5MCN-Fe was estimated to be 0.12% and 0.11%. Glucosamine hydrochloride was used as a sacrificial agent for photocatalytic seawater decomposition and using Fe-doped MCN to improve the efficiency of solar energy utilization. The prepared samples showed resident photocatalytic activity for the decomposition and hydrogen precipitation of glucosamine hydrochloride, as shown in Figure 10.7 (c). Notably, glucosamine hydrochloride as a sacrificial agent was almost completely decomposed and produced 15.84 μmol/(h·g) of hydrogen. Photo-excited charge-hole carriers have excellent optical properties and separation rates that can assume enhanced photocatalytic activity.

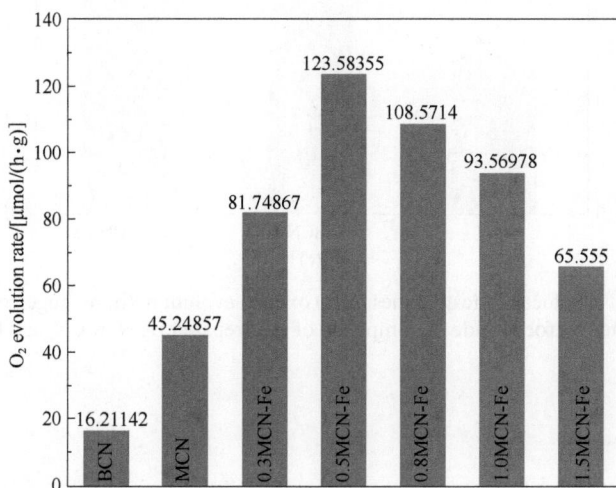

(a)

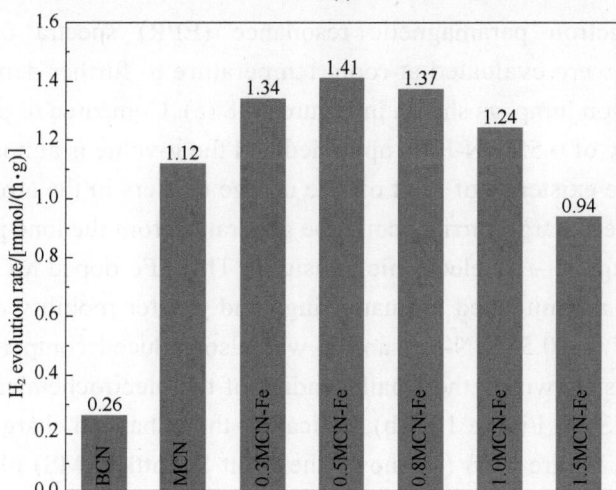

(b)

Figure 10.7

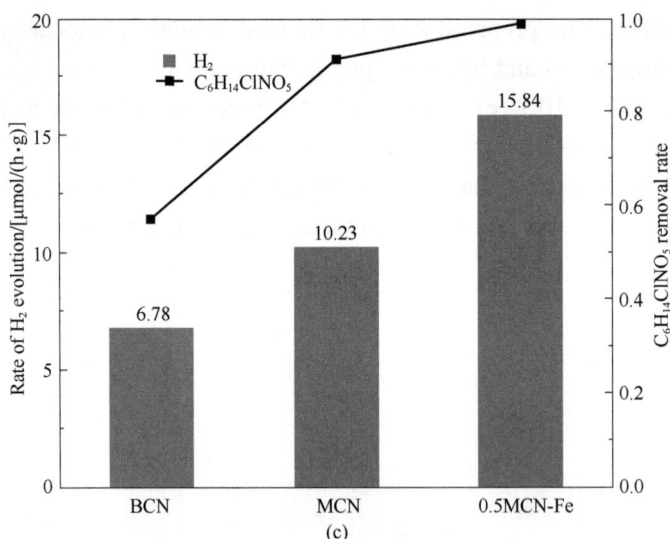

Figure 10.7   Photocatalytic activities of (a) oxygen evolution, (b) hydrogen evolution,(c) glucosamine hydrochloride decomposed of the prepared BCN, MCN, and 0.5MCN-Fe

10.4.2   EPR, DFT and charge separation and transfer performance

The electron paramagnetic resonance (EPR) spectra of BCN and 0.5MCN-Fe were evaluated at room temperature to further demonstrate the n → π* electron jump, as shown in Figure 10.8 (a). Compared to pure BCN, the EPR intensity of 0.5MCN-Fe is upgraded and the g-value is almost no change, indicating the existence of a lot of free charge carriers in the aromatic ring[11] and lots of free charge carriers could be generated from the lone pair electrons excited through n → π* electronic transition. Thus, Fe-doped MCN favors the extension of π-conjugated aromatic rings and greater mobility of spins. The resistance of the 0.5MCN-Fe sample was also reduced compared to that of pure BCN, as shown by the smaller radius of the electrochemical impedance spectrum (EIS) in Figure 10.8(b), indicating the enhanced charge mobility in 0.5MCN-Fe. Figure 10.8 (c) shows the Mott-Schottky (MS) plots for BCN and 0.5MCN-Fe. According to Figure 10.8 (c), the CB potentials of BCN and 0.5MCN-Fe were obtained from the MS plots with flat band potentials. The doped Fe element can affect the bandgap structure of MCN by tuning the electronic structure. It is obvious that the CB of 0.5MCN-Fe is more negative than that of BCN. The photogenerated electrons accumulated on 0.5MCN-Fe can be transferred to the Pt cocatalyst to generate H_2 due to the sufficiently large potential difference. The holes in the sample VB interact with the Ag co-catalyst to generate O_2. The doping of Fe atoms can modulate the electronic structure of MCN and improve the separation efficiency of photogenerated charge carriers. DFT calculations are implemented to further understand the excellent activity of MCN-Fe.

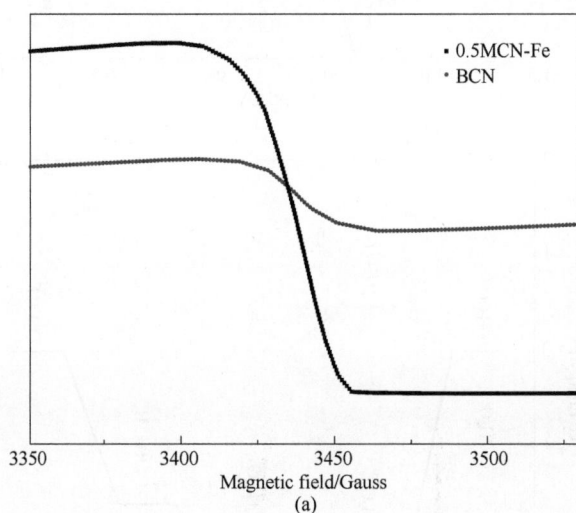

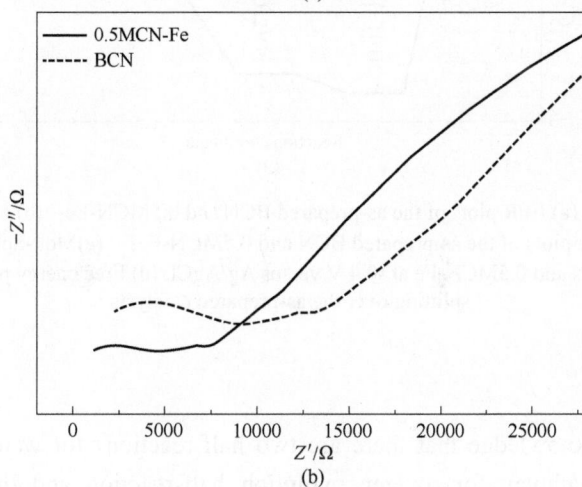

Figure 10.8

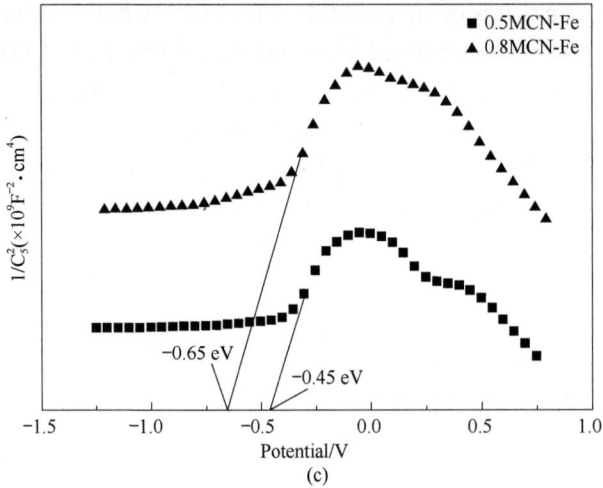

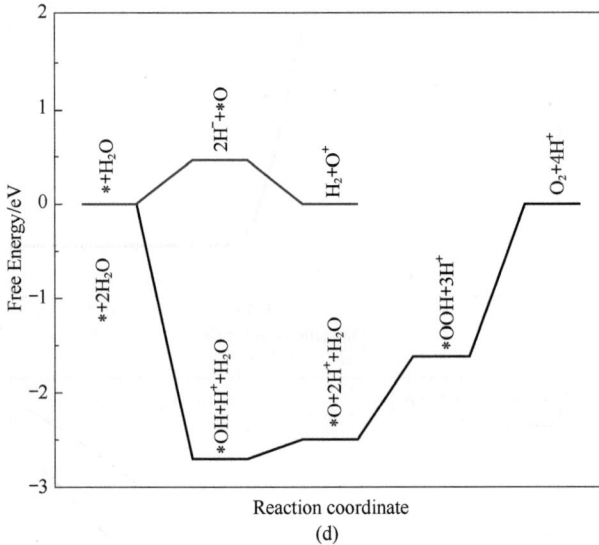

Figure 10.8 (a) EPR plots of the as-prepared BCN and 0.5MCN-Fe; (b) electrochemical impedance spectra plots of the as-prepared BCN and 0.5MCN-Fe; (c)Mott-Schottky plots of the as-prepared BCN and 0.5MCN-Fe at -0.4 V versus Ag/AgCl, (d) Free energy profiles for water splitting over the as-prepared catalysts

It is common knowledge that there are two half-reactions for water splitting, the four-electron pathway for oxygen evolution half-reaction and the two-electron pathway for hydrogen evolution half-reaction. The free energy of each pathway is calculated for the prepared MCN-Fe, as shown in Figure 10.8 (d). It can be observed that the potential limiting step of two-electron pathways of hydrogen evolution half-reaction of MCN-Fe is obtained to be the H^+ generation with a corresponding calculated overpotential of 0.467 V, while the potential limiting step of four electron pathways of oxygen evolution half-reaction is determined to be the O_2 generation with a corresponding calculated overpotential of 0.782 V. The study of BCN is mainly aimed at decreasing the compounding rate of photogenerated electron-hole pairs while reducing the energy band gap.

10.4.3　Mechanism of hydrogen production

Figure 10.9 shows the photocatalytic mechanism of MCN-Fe for water splitting and BPA degradation. The band gap effective procedure can be termed as (i) the creation of heterojunctions such as metal-semiconductors that can efficiently separate electrons and holes and (ii) metal elements doped in MCN that act as electron capture centers, leaving highly oxidized holes. It has been shown that the loading of metal ions in MCN polymers not only improves the carrier lifetime and mobility but also reduces the band gap of MCN. And (iii) the redox process of Fe and (iv) the intrinsic electron hopping ($\pi \rightarrow \pi^*$) of photoinduced electrons in MCN with π-conjugated aromatic backbones.

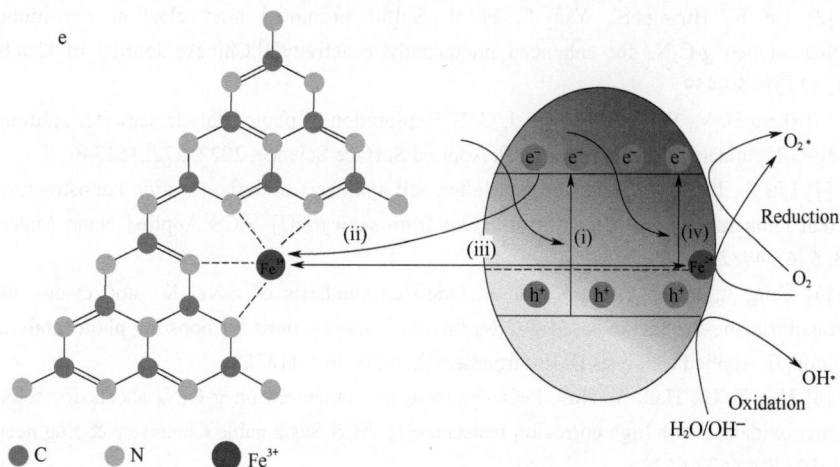

Figure 10.9　Photocatalytic mechanism of MCN-Fe for water splitting and BPA degradation

10.5 Conclusion

The preparation of Fe-doped carbon nitride as an efficient visible-light-driven photocatalyst has been demonstrated by a simple, efficient, and environmentally friendly two-step process. Based on detailed optical and photoelectrochemical characterization and DFT calculations, Fe-doped MCN provided a fast transfer of photoexcited electrons and separation rate of e^-/h^+ pairs, which enhanced the photocatalytic simulated seawater activity through the extension of the π-conjugated aromatic framework. The prepared MCN based photocatalysts had higher photocatalytic H_2 and O_2 yields than BCN. Among them, 0.5 MCN-Fe showed the best photocatalytic performance, producing 123.58 μmol O_2 and 1.41 mmol H_2 per hour. The results of the study provide new avenues for the practical application of photocatalysts.

Reference

[1] Lai Y. H., Yeh P. W., Jhong M. J., et al. Solar-driven hydrogen evolution in alkaline seawater over earth-abundant g-C_3N_4/$CuFeO_2$ heterojunction photocatalyst using microplastic as a feedstock[J]. Chemical Engineering Journal, 2023, 475: 146413.

[2] Ge F., Huang S., Yan J., et al. Sulfur promoted n-π^* electron transitions in thiophene-doped g-C_3N_4 for enhanced photocatalytic activity[J].Chinese Journal of Catalysis, 2021, 42 (3):450-459.

[3] Dang H. V., Wang Y. H., Wu J. C. S. Exploration of photocatalytic seawater splitting on Pt/GaP-C_3N_4 under simulated sunlight[J]. Applied Surface Science, 2022, 572: 151346.

[4] Liu S., Li X., Huang L.,et al. Molten salt synthesis of carbon nitride nanostructures at different temperatures for extracting uranium from seawater[J]. ACS Applied Nano Materials, 2023, 6 (6): 4782-4792.

[5] Yang J., Liang Y., Li K., et al. One-step synthesis of novel K+ and cyano groups decorated triazine-/heptazine-based g-C_3N_4 tubular homojunctions for boosting photocatalytic H_2 evolution[J]. Applied Catalysis B: Environmental, 2020, 262: 118252.

[6] Haq T. U., Haik Y. NiO_x-FeO_x Nanoclusters anchored on g-C_3N_4 sheets for selective seawater oxidation with high corrosion resistance[J]. ACS Sustainable Chemistry & Engineering, 2022, 10 (20): 6622-6632.

[7] Su C., Zhou Y., Zhang L., et al. Enhanced n$\rightarrow\pi^*$ electron transition of porous P-doped

g-C$_3$N$_4$ nanosheets for improved photocatalytic H$_2$ evolution performance[J]. Ceramics International, 2020, 46 (6): 8444-8451.

[8] Liu X., Yan L., Hu X., et al. Facile synthesis of B and P doped g-C$_3$N$_4$ for enhanced synergetic activity between photocatalytic water splitting and BPA degradation[J]. International Journal of Hydrogen Energy, 2023.

[9] Jin D., He D., Lv Y., et al. Preparation of metal-free BP/CN photocatalyst with enhanced ability for photocatalytic tetracycline degradation[J]. Chemosphere, 2022, 290: 133317.

[10] Liu X., Mateen M., Cheng X., et al. Constructing atomic Co^1-N$_4$ sites in 2D polymeric carbon nitride for boosting photocatalytic hydrogen harvesting under visible light[J]. International Journal of Hydrogen Energy, 2022, 47 (25): 12592-12604.

[11] Zhang G., Li G., Heil T., et al. Tailoring the grain boundary chemistry of polymeric carbon nitride for enhanced solar hydrogen production and CO$_{(2)}$ reduction[J]. Angewandte Chemie (International ed. in English), 2019, 58 (11): 3433-3437.